W9-BUH-797

THE FIRST
20
MINUTES

THE FIRST
20
MINUTES

SURPRISING SCIENCE REVEALS
HOW WE CAN EXERCISE BETTER,
TRAIN SMARTER, LIVE LONGER

Gretchen Reynolds

HUDSON
STREET
PRESS

HUDSON STREET PRESS
Published by the Penguin Group
Penguin Group (USA) Inc., 375 Hudson Street, New York, New York 10014, U.S.A. • Penguin Group (Canada), 90 Eglinton Avenue East, Suite 700, Toronto, Ontario, Canada M4P 2Y3 (a division of Pearson Penguin Canada Inc.) • Penguin Books Ltd., 80 Strand, London WC2R 0RL, England • Penguin Ireland, 25 St. Stephen's Green, Dublin 2, Ireland (a division of Penguin Books Ltd.) • Penguin Group (Australia), 250 Camberwell Road, Camberwell, Victoria 3124, Australia (a division of Pearson Australia Group Pty. Ltd.) • Penguin Books India Pvt. Ltd., 11 Community Centre, Panchsheel Park, New Delhi – 110 017, India • Penguin Group (NZ), 67 Apollo Drive, Rosedale, Auckland 0632, New Zealand (a division of Pearson New Zealand Ltd.) • Penguin Books (South Africa) (Pty.) Ltd., 24 Sturdee Avenue, Rosebank, Johannesburg 2196, South Africa

Penguin Books Ltd., Registered Offices: 80 Strand, London WC2R 0RL, England

First published by Hudson Street Press, a member of Penguin Group (USA) Inc.

First Printing, May 2012
10 9 8 7 6 5 4 3

REGISTERED TRADEMARK—MARCA REGISTRADA

HUDSON
STREET
PRESS

LIBRARY OF CONGRESS CATALOGING-IN-PUBLICATION DATA

Reynolds, Gretchen.
The first 20 minutes : the myth-busting science that shows how we can walk farther, run faster, and live longer / Gretchen Reynolds.
    p. cm.
Includes index.
ISBN 978-1-59463-093-4 (alk. paper)
    1. Physical fitness. 2. Physical education and training—Physiological aspects. 3. Exercise—Physiological aspects. I. Title. II. Title: First twenty minutes.
GV481.R49 2012
613.7—dc23
                              2012000321

Printed in the United States of America
Set in Adobe Garamond Pro

PUBLISHER'S NOTE
Every effort has been made to ensure that the information contained in this book is complete and accurate. However, neither the publisher nor the author is engaged in rendering professional advice or services to the individual reader. The ideas, procedures, and suggestions contained in this book are not intended as a substitute for consulting with your physician. All matters regarding your health require medical supervision. Neither the author nor the publisher shall be liable or responsible for any loss or damage allegedly arising from any information or suggestion in this book.

To Russell and Max,
my training partners for life

A body in motion remains in motion, unless acted upon by an external force. A body at rest remains at rest.

—Newton's First Law of Motion

Eating alone will not keep a man well; he must also take exercise.

—Hippocrates

CONTENTS

ACKNOWLEDGMENTS

This book wouldn't have been possible without the help of many people, especially the scientists who performed the experiments that I write about. Theirs is the hard and groundbreaking work, and I thank them wholeheartedly for patiently answering my many, basic, and sometimes inane questions. I particularly want to acknowledge the assistance and expertise of Dr. Dan Carey, who passed away while I was still working on the book. Thanks, too, to Mark Bryant, the founding editor of *Play* magazine at the *New York Times*, who suggested that I write a health and fitness column for the magazine; and to my other editors at the *Times*: Ilena Silverman, Gerald Marzorati, Laura Hohnhold, Tara Parker-Pope, and Toby Bilanow. Special thanks to my agent Sam Stoloff and my editor Meghan Stevenson, who kept the chapters flowing when I was afraid the spigot was dry. Finally, warm thanks to my family, by blood and marriage, who put up with my long hours, frequent absences, and whining.

A Body in Motion

It's become a truism to say that humankind was born to run, and perhaps some politicians were, though what proper and odd little children they must have been. For the rest of us, the notion is debatable. Consider the latest evidence. Not long ago, biologists from the University of Utah, together with zoologists from Friedrich Schiller University Jena, in Germany, and other researchers set out to study how humans move, and how divergent our locomotion is from that of, say, cheetahs or gazelles, who run so fluidly. The researchers fitted volunteers, most of them trained athletes in their twenties, thirties, or forties, with face masks that measured how much oxygen they were breathing and then had them alternately walk or run on treadmills.

Each of the volunteers was asked to run and to walk in three different ways: by landing either on their heels, on the middle of their feet, or on their toes. Some research in the past few years has suggested that humans run most efficiently, meaning with the least consumption of oxygen, when they land on the balls of their feet or on their toes. Efficiency in human movement is desirable, as it is in other forms of transport, because it means that you burn less fuel over a given distance. Like a fuel-efficient car, a more efficient human machine needs less energy to cover the same number of miles as an inefficient one. Fuel savings matters during evolution. Being able to amble long distances on less food than the next *Homo sapiens* could mean that you'd complete a long, stalking hunt and still have the wherewithal to undertake carnal relations later and pass your DNA on to the next generation. Part of the thinking behind the idea that humans are born to run is that running, particularly if you land on your toes, should be an efficient stride and that, employing it, we would have run easily for hours, chasing down our prey.

But in this study, running was not the most efficient human stride, not by a long shot. It didn't matter whether a runner landed on his toes, the balls of his feet, or his heels. Running just wasn't fuel efficient, the data showed. Walking was. By a sizeable margin, walking, especially when the athletes landed first with their heels, was the most physically economical way for human beings to move. This estranges us from much of the animal world. Gazelles rarely walk and don't do it well. They bound madly, landing on their toes. But humans seem built to plod. "We are remarkably economical walkers," the authors concluded. "We are not efficient runners. We consume more energy to run than the typical mammal our size."

Conveniently, as it turns out, caveman-like hunts were probably conducted at a walking speed, anyway. When researchers recently followed a group of modern-day African hunters on a long, slow pursuit of their prey, the average speed was 3.8 miles per hour, a walking pace.

"This notion that all humans were born to run is unscientific," says zoologist Karen Steudel, Ph.D., of the University of Wisconsin-Madison, who's conducted a number of comparative studies of the evolution of human and animal locomotion. "The evolutionary record makes it clear that humans are born to be active," she adds. Sitting in one place wasn't an effective survival strategy when big cats and mammoths were around and food was mobile. "That's all we know for sure at the moment," she continues. "But ideas can change with new discoveries. Check back in a month."

This is a book about your body in motion. It's also a book about change, because what we are learning today about the moving body is itself a flying target. Exercise science has never been so yeasty. Every week brings a new discovery that undercuts another entrenched (and often beloved) exercise practice. Who once would have believed that massage would turn out not to help tired muscles to recover? Or that chocolate milk would? For years we were told it was impossible to drink too much water during a marathon, but overdrinking, it's been proven, can kill. The litany goes on. Stretching is probably bad for your muscles, but running is good for your knees. Weight training makes you smart. Lucky underpants really work. Your genes might be the reason you're so reluctant to work out in the first place. And humans are born to stroll.

It's a fascinating time to own a body, and a perplexing one.

But, really, we shouldn't be surprised that our understanding of the moving body is in flux. The body is an astonishingly complicated contraption. The machinations required just to lift your finger are boggling. Muscles interlace with nerves, tendons, ligaments, and bones. Collagen stretches against sarcomeres. Cartilage softens the rubbing of bones. The brain, initiating movement, is flooded with spidery feedback from muscles, eyes, skin, various of its own lobes, and other systems. Fluids move in and out of cells. Biochemical processes flare. The liver gets drawn in.

For many years, the specifics of this process were baffling. Scientists simply didn't have the tools to determine some aspects of how the exercising body worked. Organs and bodily systems were inaccessible or inconvenient for study purposes.

But now, with the rapid advancements in microscopy, neurology, radiology, surgery, 3-D biomechanical imaging, and other fields, we can see into and measure the human body as never before. Consider what we've learned just about the brain in recent years.

Sitting behind a dense, bony, protective skull, it didn't, with standard imaging equipment, seem to be doing much, no matter how hard someone was thinking or how elaborately he was moving. But with the advent of functional magnetic resonance imaging (fMRI) machines, scientists can view, from outside, the brain's operations—which portions are firing with electrical impulses during movement and which portions are remodeled *by* movement. Because, make no mistake, as the latest science absolutely assures us, no part of your body is static. If you move enough, your muscles change and grow. So does your mind. The brain initiates movement. But it is, in its turn, remade by movement. New cells are born; new vessels sprout. The same process operates body-wide. No cell in your body is unaffected by motion. Your very DNA is changed.

So, move. The state-of-the-art exercise science also points out, with increasing urgency, that inactivity is, for the human body, unnatural and unwise. Death rates rise when societies sit. Waistlines grow. Unhappiness spreads. By one recent calculation, life expectancy in the United States and Europe will drop in the coming decade, for the first time in modern history, in large part because people do not exercise enough (but also because we smoke and overeat).

So, if this book is partly an overview of what's known right now about physical performance, it's also—and perhaps even more accurately—a user's manual. I hope it will allow you to take what's being learned about the human body and put it into action, whether your aim is to break 3:30 in a marathon or to walk more briskly around the block. We all have questions about exercise, whether we've been working out for years or hope to start tomorrow. Do you have to follow a specific diet? Can you get fit with only a few minutes of exercise on any given day? What is an interval? Science is, with exhausting regularity, supplying new, tested, evidence-based answers to these and hundreds of other questions.

We don't, after all, have to be athletes to want to know how best to move. We need only to listen to the voice bred deep into our blood and bones that says, "Hey, let's go for a walk. The antelope herds are moving." (You can ignore that last part.) The body wants to move. Go with it. And the information about intervals, by the way, can be found on page 12.

THE FIRST
20
MINUTES

1

The First 20 Minutes

Do you have many aches and pains? Is your mood generally good? How much time do you spend jogging? Do you ever visit the gym? Periodically, the good folks at the Division of Adult and Community Health, at the Centers for Disease Control and Prevention, in Atlanta, check in with Americans to see how they're feeling about their health, asking them such questions as part of the ambitious and sweeping telephone poll known as the Behavioral Risk Factor Surveillance System or, more familiarly, the BRFSS (gesundheit!). This survey asks Americans about their physical activities, including whether they engage in any exercise, and about how they subjectively feel about their "health-related quality of life."

After the raw data from one of the recent BRFSSs became available, CDC researchers decided for the first time to cross-correlate the information about, on the one hand, people's activity levels and, on the other, their health-related quality of life, on a monthly basis. The researchers had anticipated, as they wrote in their published report, that "physical activity" would be "associated with increasing benefits to health," both physiological and emotional, although as they also recognized, the "dose-response relationships between physical activity and many health benefits remains unclear." In other words, the researchers felt confident that exercise was good for you, but they weren't quite sure how much was necessary to receive benefits.

Their report, published in *Medicine & Science in Sports & Exercise*, the official journal of the American College of Sports Medicine, somehow managed to muddy the issue further. They found that of the 175,850 adults whose health information was parsed, 18 percent engaged in effectively zero planned physical activity (i.e. exercise), while 66 percent completed at least 30 minutes a day of moderate physical activity (such as walking or easy bicycling) and 42 percent said they exercised vigorously (jogging, for instance) at least once or twice a week for 20 minutes or more. (Many in this group were also moderate exercisers on other days.) These are more impressive numbers, in terms of activity, than in many recent studies of Americans. In those, particularly when the studies relied on hard measurements, such as pedometers, to gauge activity, the percentage of Americans who were even moderately active on most days of the week barely reached 50 percent.

The more eye-opening BRFSS data, though, came from people's estimates of their health-related quality of life in the month preceding the survey. People who exercised moderately reported fewer "unhealthy" days, during which they felt fatigued, unhappy, ill, anxious, achy, or otherwise "off," than people who didn't exercise. Almost 30 percent of the sedentary respondents, in fact, said they'd felt puny on at least 14 days in the prior month. Far more surprising, though, was that more than 20 percent of the people who said they worked out vigorously multiple times during the week also reported 14 or more "unhealthy days" in the month. Specifically, since this was a study overseen by health statisticians, the scientists wrote that "a poor HRQOL [health-related quality of life] was always more likely among those with no physical activity, usually more likely among those who had daily (7 days a week) activity, almost always more likely for those with activity of short duration (less than 20 minutes a day) and more likely more than half the time for those with very long duration (more than 90 minutes a day)."

Or, to be blunt, the issue of just how much exercise people need and how much may be either too little or too much is, from a scientific standpoint, a big fat mess.

Not Stepping Up

There was a time when the question of how much exercise a person required was moot. The cows needed seeing to; the corn needed tending. As we all know, prior to World War II, most Americans lived outside cities and were active almost all the time, whether they wished to be or not. A recent study of activity levels among a group of modern Old Order Amish families, whose lifestyles are considered representative of a past America (apart from the boomers), found that Amish men spent more than 10 hours a week in vigorous activity, on top of almost 43 hours a week of moderate activity and 12 hours a week of walking. They averaged almost 18,500 steps per day, or about nine miles of walking every day of the week except Sunday. The Amish women were relatively slothful, covering only about 7.5 miles per day, on average.

By comparison, according to 2010 statistics, most American adults take about 5,000 steps a day, which pales in comparison not only with the Amish but also with activity levels in other countries. The happy-go-lucky Australians average about 9,700 steps a day, the highest total in the Western world. The Swiss, number two, yodel through 9,650 steps a day and, despite the ready availability of Lindt chocolate, have a national obesity rate of barely 8 percent. In America, that rate is 34 percent and rising.

But while those figures make it clear that most Americans don't move enough, they don't tell us how much each of us should be moving, because, frankly, no one really knows. "Science and common sense tell us that, without a doubt, it's unhealthy to sit and be sedentary all day," says William Haskell, Ph.D., an emeritus professor of exercise physiology at Stanford University and one of the country's experts in exercise dosing and longevity. "But precisely how much exercise is required for health, fitness, or athletic performance is difficult to determine."

Health, fitness, and athletic performance are, after all, distinct aims with distinct demands, and each of us must resolve, for ourselves, which we're trying to achieve. We also must decide how much we're willing to do, realistically, to reach those standards. Health may seem the most achievable goal, but in reality *health* is a slippery term, defined often by its absence. Having high blood pressure, rotten cholesterol numbers, too much blood sugar, a wide waist, or actual illnesses, from colds to cancer, is un-healthy. Not experiencing those same conditions is good health. Activity can, if chosen wisely, improve health.

Fitness is something else, although health and fitness are often automatically joined together. If you ask an exercise physiologist, *fitness* refers to cardiovascular or cardiorespiratory fitness (the two terms are almost but not quite synonymous—cardiorespiratory includes measures of lung function—but close enough). Physical fitness in this sense is a measure of how efficiently you transport oxygen to laboring muscles and maintain movement. A physically fit person has strong lungs, a robust heart, and sturdy muscles. She may or may not be clinically healthy. Some people blessed with high marks on fitness can have miserable cholesterol profiles or rotund waistlines. A surprisingly large portion of any given person's biological fitness is, in fact, innate. According to several large recent studies, 30 percent or more of a person's cardiovascular fitness may be genetic. You are born either more or less physically fit than the next person. But how you augment or diminish that inheritance is up to you.

Finally, there's athletic performance, an ambition unto itself, capable, in some instances, of mitigating the other two. Walking three miles a day on a regular basis will almost certainly improve most people's health and fitness. Running four marathons in a year might not. Unless it does. "There is considerable variability in people's responses to exercise, at any dose," Dr. Haskell says.

Which raises the most central and pressing question in this entire book: Yes, fine, all those studies are very interesting; but what about *me*?

How Low Can You Go?

Recently researchers in Scotland trawled through a vast database of survey data about the health and habits of men and women in that fair land, similar to the BRFSS survey here. In this case, the scientists were looking to see how much exercise was needed to keep the average Scotsman or -woman from feeling dour (or in technical terms, experiencing "psychological distress"). Scots are not famed for being blithe-hearted, and many of us might have expected that firm measures and lots of sweat would be required. But as it turned out, researchers found that a mere twenty minutes a week—a week!—of any physical activity, whether vigorous or easy, improved the respondents' dispositions. The activities in question ranged from organized sports to walking, gardening, and even housecleaning, the last not usually associated with bliss. The researchers found that, in general, more activity did confer more mental-health benefits and that "participation in vigorous sports activities" tended to be the "most beneficial for mental health." But overall their conclusion was that being active for as little as twenty minutes a week was sufficient, if your specific goal happened to be a sanguine temper.

The question of just how little activity people can get away with has preoccupied exercise scientists in recent years, in part because so many of us have proven so resistant to any exercise. There was a time, in the 1970s and 1980s, when most exercise guidelines, including those from the American College of Sports Medicine (ACSM) and other groups, aimed at athleticism; they recommended that people engage in prolonged, uninterrupted, vigorous activity for an hour or more, multiple times a week. Basically, people should run, swim, or bicycle, the recommendations suggested, and they should do so hard, and the more the better.

Some people responded. That was the height of the 1970s running boom. Then in 1984, Jim Fixx, the author of *The Complete Book of Running*, died at age fifty-two of a "fulminant heart attack" while marathon training. Running didn't kill him. He'd been afflicted, an autopsy showed, with intractable heart disease, probably congenital. But some people gleefully and ghoulishly pointed to his death as a reason to remain couch-bound. Even more Americans, though, hadn't needed such an excuse. They had not been inspired to exercise in the first place, at least not hard, and resolutely continued not to.

By the 1990s, formal exercise recommendations, bowing to human nature, had softened, and experts were suggesting that less-vigorous exercise might be sufficient. In 1995, the ACSM and the CDC jointly announced that "Every U.S. adult should accumulate thirty minutes or more of moderate-intensity physical activity on most, preferably all, days of the week."

But there still was little science behind any exercise guidelines, including that one. So in the mid-2000s, the U.S. Department of Health and Human Services convened an advisory committee of scientists, including physiologists, cardiologists, epidemiologists, nutritionists, and others, and asked them to scrutinize decades of studies about the benefits—and risks—of exercise and to formulate new, evidence-based guidelines. The result was the massive 2008 *Physical Activity Guidelines for Americans*, which began on a cautionary note. The "amount of physical activity necessary to produce health benefits cannot yet be identified with a high degree of precision," the authors wrote.

Oh. Great.

But still the scientists had persevered, wading through studies involving animals and humans, looking at the impacts that various types and amounts of activity have on people's risks for heart disease, obesity, diabetes, cancer, depression, and premature death in general. In some studies cited in the final report, exercise conferred little if any disease-fighting benefits. In others, the benefits kicked in only if the exercise was quite strenuous. In still others, a gentle stroll a few times a week was enough to lessen the risk of many diseases and early death.

Despite the inconsistent results, the advisory committee ultimately reached a consensus about just how much—and how little— exercise most of us should be getting, at least for health purposes. The magic number, the report announced, was a minimum of 500 MET minutes of exercise a week.

So, get to it.

Of course, unless you're an exercise scientist, you probably don't know what a MET minute is. A single MET, or Metabolic Equivalent of Task, is the amount of energy a person uses at rest. Two METs represent twice the energy burned at rest; four METs, four times the energy used at rest; and so on. Walking at three miles per hour is a 3.3-MET activity, while running at 6 miles per hour is a 10-MET activity. The committee concluded that a person needs to accumulate a weekly minimum of 500 MET minutes of exercise, which does not mean 500 minutes of exercise. Instead, 150 minutes a week (two and a half hours) of a moderate, three- to five-MET activity, such as walking, works out to be about 500 MET minutes. Half as much time (an hour and 15 minutes per week) spent on a 6-plus MET activity such as easy jogging seems, according to the committee, to have similar health effects.

What this means, in practical terms, is that according to the best available science, you should walk or otherwise work out lightly for 150 minutes a week in order to improve your health. This report and other, newer science show that you can split these 150 minutes into almost any chunks and still benefit. In a nifty study of aerospace engineers (virgin exercisers, one and all), the men were assigned to briskly walk or gently jog for 30 minutes a day in either a single, uninterrupted half-hour bout or in three 10-minute sessions spread throughout the day (10 minutes in the morning, 10 minutes at lunchtime, and 10 minutes in the evening). At the end of eight weeks, both groups of engineers had improved their health and fitness profiles, without major differences between the groups. All had wound up with lower heart rates, better endurance on a treadmill test, and a few less pounds.

So, too, when separate groups of lab rats recently were allowed either leisurely to rodent-paddle in a laboratory water feature for three hours or were required to increase the tempo until they were swimming quite vigorously for 45 minutes, the animals afterward uniformly displayed significant improvements in their bodies' ability to regulate blood sugar, a key measure of health.

It hadn't mattered how they'd accumulated the exercise, only that they had.

The Overload Principle

There is a catch. The person who is likely to benefit the most from increasing exercise time is probably not you, but instead your pudgy uncle Clarence or that pasty kid next door who's never met an online orc he couldn't slay. "The greatest health benefit from exercise comes from getting up off the couch," says exercise physiologist Timothy Church, Ph.D., a professor at the Pennington Biomedical Research Center, in Baton Rouge, Louisiana, who has studied exercise dosing extensively. "Everything after that is incremental."

The health benefits of activity follow, in fact, a breathtakingly steep curve at first. "Almost all of the mortality reductions are due to the first twenty minutes of exercise," says Frank Booth, Ph.D., a professor of biomedical sciences at the University of Missouri and much-cited expert on exercise and health. "There's a huge drop in mortality rates among people who haven't been doing any activity and then begin doing some, even if the amount of exercise is quite small." In a recent meta-analysis of studies about exercise and mortality conducted by scientists at the University of Cambridge, in England, and others, the authors found that in general a person's risk of dying prematurely from any cause plummeted by nearly 20 percent if he or she began to meet the current exercise guidelines of 150 minutes of moderate activity per week, compared with someone who didn't exercise.

If, however, someone almost tripled that minimum level, completing about ninety minutes a day of exercise four or five times a week, the researchers wrote, his or her risk of premature death dropped still further, but only by another 4 percent.

He or she might, however, be well on the way to enviable fitness. That's important to bear in mind, especially for those, like me, who are fairly confident that we sit outside the bounds of mortality and don't need to obsess about premature death but really, really want to trounce our spouses at the next community Fun Run. Any amount and type of exercise will probably improve your health somewhat, but it won't necessarily make you physically fitter or athletically more competitive. Thirty minutes of walking five times a week is not going to lower your personal best time in a 5K.

Walking three minutes at an extremely brisk pace, though, followed by three minutes of slower striding, with that set repeated five or six times, just might. There is a famous fitness principle known as "overload," which, according to a recent commentary in the well-regarded *Journal of Applied Physiology*, is "the one overriding truth in exercise physiology." Encouraging, isn't it, that there is at least one?

Overload is not a complicated idea. The word encapsulates the concept. *Overload* simply means that, as the commentary explains it, "improved athletic performance is the result of systematic and progressive training of sufficient frequency, intensity, and duration." You can't keep on doing the same old workout and improve athletically. The body gets used to a certain level of activity with impressive rapidity. So you have to ratchet things up.

You've no doubt experienced overload in action. Maybe you used to puff and struggle on the elliptical machine after twenty minutes and soon felt obliged to quit for the day. Then after a few weeks those same twenty minutes became easy. From then on, you could, should you so choose, repeat that same undemanding workout—with unchanged time, distance, and resistance level—for the rest of your life and continue to accrue health benefits.

But if you wanted to become fitter, faster, or in general tougher, you'd have to dial up the resistance or prolong the workout. You'd puff and struggle again, and slowly grow used to the new workload. You would have overloaded your cardiovascular and other systems, let them readjust, and from a fitness and athletic standpoint, improved. Good for you.

But how? The practical options are, you may be pleased to know, many. It's never been difficult to overload oneself in this modern world. Google Calendar can do it for you. But physiologically, you achieve overload by increasing the number of times you work out in a week, the length of time each workout lasts, or the intensity of any given workout. If you enjoy walking and currently schedule thirty-minute walks five times a week, you can lengthen each session to thirty-five minutes, then forty. As a general rule, you shouldn't increase your training volume by much more than 10 percent a week, to avoid injury. But exercise of low intensity, such as walking, rarely results in injury in the first place.

You also can increase the intensity of the same workout, a concept that many of us know and dread as intervals. Intervals are typically short, sharp, unpleasant bouts of exercise performed at an intensity as close to your maximum as you can stand, followed by a rest period and then repeated. They definitely result in overload and occasionally nausea among athletes, many of whom pepper several interval sessions into a week's worth of competitive training.

The latest science suggests, though, that benign versions of interval training can provide significant performance benefits even to walkers, recreational athletes, or any of us who wish to improve athletically but not to vomit. In a cheering, ongoing experiment in Japan, middle-aged and elderly walkers were assigned to one of two programs. Some undertook a low-intensity regimen, during which they strode eight thousand steps (or between three and four miles, as measured by pedometers) at an intensity equal to about 50 percent of their maximum heart rates—nice, easy walking. The rest practiced intervals, which consisted in their case of three-minute bursts of power walking, during which they'd raise their heart rates to about 70 percent of their predetermined maximum. This was followed by three minutes of striding at barely 40 percent of their maximum heart rates, then three minutes hard again, and so on, until they'd completed at least five sets.

Both groups walked for five months. At the end, all of the walkers had improved their blood pressure readings. But only the interval group had developed measurably greater leg muscle power, as determined by weight-training machines, and higher maximum oxygen capacities. They had grown significantly more physically fit than the strollers. The intervals had done their job.

The regimen can be so effective, in fact, that it has the potential to provide all of the exercise that you need in only a few minutes per week. Those minutes, though, will not be pleasant.

The Four-and-a-Half-Minute Workout!

In a laboratory at the National Institute of Health and Nutrition, in Japan, scientists watched with interest as laboratory rats paddled furiously around the periphery of a shallow barrel filled with warm water. There was one barrel for every rat. The swimmers had been at it for hours. Swimming is a reliable way for scientists to test and increase rodents' aerobic fitness, because most rats aren't very good at it. There is no placid backstroke for a rat. It splashes and slaps and uses plenty of energy. A prolonged session of paddling can tax a rat.

After three hours, the scientists scooped the rats out of the water and let them rest quietly for forty-five minutes. Then they lowered the rats back into the barrel and had them swim for an additional three hours. Afterward, the researchers tested each rat's leg-muscle fibers and found that they had begun to show biochemical, molecular changes that indicated that the rats were increasing their bodies' endurance and fitness.

Meanwhile, a second group of rats was motoring through vastly shorter swimming sessions. These rats, wearing tiny weighted vests equivalent to 14 percent of their body weight (to make the swimming more strenuous), windmilled around the barrels for 20 seconds and were then lifted out and allowed to rest for 10 seconds. They completed 14 of these vigorous mini-swims, for a grand total of about 4.5 minutes of swimming. When the scientists biopsied their muscle fibers, they found the same molecular changes as in the long-distance swimmers, and more of them. Four and a half minutes of intense exercise had yielded, it would seem, virtually the same aerobic benefits as six long hours in the water.

The concept of "high-intensity interval training," or HIIT, is relatively new and quite different from the old-school approach to intervals that most of us remember from high school track. With HIT, you don't intersperse interval sessions on one day with longer workouts on others. You only do intervals, day after day, finishing the hard work in a matter of minutes. "There was a time when the scientific literature suggested that the only way to achieve endurance was through endurance-type activities," such as long, relatively easy runs or bike rides or, perhaps, six-hour swims, says Martin Gibala, Ph.D., a professor at McMaster University in Canada, who's been at the forefront of HIIT science.

But ongoing research from Dr. Gibala's lab proves otherwise. In one study (which was the most e-mailed document on the website of the *Journal of Applied Physiology* for almost two years), Dr. Gibala and his colleagues had a group of healthy college students ride a stationary bike at a sustainable pace for between 90 and 120 minutes. Another set of students grunted through a series of short, strenuous intervals: 20 to 30 seconds of cycling at the highest intensity the riders could stand. "We describe it as an 'all-out' effort," Dr. Gibala says, which requires straying "well out of your comfort zone." After resting for 4 minutes, the students pedaled hard again for another 20 to 30 seconds, repeating the cycle 4 to 6 times (depending on how much each person could stand), "for a total of two to three minutes of very intense exercise per training session," Dr. Gibala says.

Each of the two groups exercised three times a week. After two weeks, both groups showed almost identical increases in their endurance (as measured in a stationary bicycle time trial), even though the one group had exercised for six to nine minutes per week, and the other for about five hours each week. Both groups, in biopsies, showed dramatic molecular changes deep within their muscle cells indicating increased physical fitness. In particular, they had far more mitochondria now, the microscopic organelles that allow muscles to use oxygen to create energy. Six minutes or so a week of hard exercise (plus the time spent warming up, cooling down, and resting between the bouts of intense work) had proven to be as good as about three hundred minutes of less strenuous exercise for achieving basic fitness.

Sadly, those six minutes had to hurt.

Jogging Trumps Berry Picking

The health benefits of vigorous exercise can be substantial, though, if that's sufficient compensation. Researchers in Finland, for instance, recently concluded that if you wish to ward off lung or gastrointestinal cancer, you should spend your leisure time jogging instead of picking berries, gathering mushrooms, or fishing. They based that finding on extraordinarily detailed health diaries from a group of 2,560 middle-aged Finnish men who have been keeping records of their daily activities for the past two decades and who reside in a nation where berry picking and mushroom hunting regularly occur.

When the study began, none of the men had cancer. Seventeen years later, 181 had died of the disease. Controlling for cigarette smoking, fiber and fat intake, age, and other obvious health-related variables, the researchers determined that activity had significantly affected the men's cancer risk. The most physically active were the least likely to develop or die from malignancies, particularly of the gastrointestinal tract or the lungs.

And intensity was key. More arduous exercise was more protective. Jogging was the most strenuous activity studied; fishing among the least. The men who jogged or otherwise exercised fairly intensely for at least thirty minutes a day had "a fifty percent reduction in the risk of dying prematurely from cancer," says Sudhir Kurl, M.D., medical director of the School of Public Health and Clinical Nutrition at Finland's University of Kuopio and one of the study's authors.

More technically, the men whose METs reached at least five almost every day were the least likely to die of cancer, especially of the lung or the gastrointestinal tract. That result jibes neatly with the findings of a large study of women and colon cancer. In it, women who walked briskly for five to six hours a week were much less likely to develop the disease than those who strolled for thirty minutes per week. And in the bogglingly comprehensive 2008 national Physical Activity Guidelines Advisory Committee report prepared for the secretary of health and human services, which includes a chapter about exercise and cancer, the authors concluded that "one hour per day of moderate or vigorous activity produces greater reduction in risk" than the two and a half hours of moderate exercise per week that is currently recommended.

The Finnish researchers admit that they don't know just how or why brisk exercise reduces cancer risk or why only some types of cancer seem to be affected. Exercise speeds the emptying of the colon, which may reduce the amount of time that carcinogens linger in the organ. Strenuous and frequent exercise also can affect the production of sex hormones in men and women, which could have implications for breast and prostate cancer. Even the panting involved in hearty exercise might somehow move carcinogens out of the lungs, the researchers point out.

But it remains difficult to tease out the specific molecular effects of regular, taxing exercise from the generally healthy habits of exercisers. "Lifestyle factors" and the luck (good and bad) of genetics could be skewing the results, the authors admit. Still, there is food for thought in their findings. "At least moderately intense physical activity is more beneficial than low intensity physical activity in the prevention of cancer," the authors conclude. Forget the romantic imagery. For this purpose, jogging trumps berry picking.

Too Much Much-ness?

Only you can decide, though, how much discomfort you wish to endure to become healthy and fit or how little time you can devote to the process. Only you can decide if fitness, with or without competition, is ultimately your goal.

And it's always possible to do too much. "The benefits of exercise appear to be curvilinear," Dr. Booth says. The profits rise precipitously when you first wade into exercise, level off as you do more, and, at some point, drop if you overdo things, although just where this break point occurs is "likely to differ from person to person," Dr. Booth says. Overuse injuries are the most common symptom that you're overdoing it. They are, after all, *overuse* injuries, which develop from accumulating wear and tear. (Acute injuries result from sudden, one-off accidents or incidents such as twisting a knee during soccer, treading in a hole while running, or clobbering yourself with a stop sign while reading a book walking down the street, and that only happened to me once.) A study of five thousand adults enrolled in the long-term Aerobics Center Longitudinal Study overseen by the Cooper Institute, in Dallas, found, unsurprisingly, that the more people exercised, the more likely they were to wind up hurt. That sounds like "duh" science, to be sure, but the conclusion is often ignored by avid athletes, who'll continue through recurrent, nagging muscle twinges until they develop a severe injury.

Or they may exercise themselves into a sickbed, as a provocative series of recent studies implies. In one, researchers divided lab mice into two groups. The first group rested comfortably in their cages. The others ran on little treadmills until they were exhausted. This continued for three days. The mice were then exposed to a flu virus. After a few days, more of the mice that had exhausted themselves running came down with the flu than the control mice. They also had more severe symptoms.

Likewise, when other scientists infected mice with a particularly virulent type of mouse flu and then had them either rest, jog slowly for twenty or thirty minutes, or run flat-out on a treadmill for two and a half hours, repeating the sessions for the next three days, the results were telling. The running mice were hardest hit by the flu. About 70 percent of them died, while only half of the sedentary mice expired, and a mere 12 percent of the gently jogging mice passed away. Meanwhile, inside the running mice, scientists found evidence of significant perturbations in their immune systems. In both mice and men, viruses can evoke an increase in immune cells that induces inflammation, a bodily defense. But if the inflammation continues for too long, it becomes counterproductive. In the running mice, cells that would normally have eased the inflammation somehow had been muted and didn't respond fully. The inflammatory process flamed out of control. The mice died.

Moderate exercise, on the other hand, had improved the mice's inflammatory cell balance. They were the healthiest and fittest animals in the lab.

Broken Hearts

At the extremes of exercise, then, there can be undesirable consequences. And among true outliers, those consequences can involve the heart. It is, after all, a muscle and, as such, potentially prone to overuse. Recently, British researchers looked closely at the heart health of a group of lifelong, seriously competitive male athletes. All of their subjects had been or were British national team distance runners or rowers, or were members of the extremely selective 100 Marathon Club, which admits runners who, as you might have guessed, have completed at least a hundred marathons. The men had trained and competed throughout their adult lives and still worked out strenuously. About half were past age 50 and the rest were relative striplings, between 26 and 40. The scientists also gathered a group of 20 healthy men over 50, none of them endurance athletes, for comparison.

The different groups underwent a sophisticated new form of magnetic resonance imaging of their hearts that can spot very early signs of fibrosis, or scarring, within the heart muscle. Fibrosis, if it becomes severe, often leads to stiffening or thickening of portions of the heart, contributing to irregular heart function and, in the worst cases, heart failure.

None of the younger athletes or the older nonathletes had fibrosis in their hearts. But half of the older lifelong athletes showed some heart muscle scarring. The affected men were, in each instance, those who'd trained the longest and hardest. None were experiencing any symptoms. But spending more years exercising strenuously was associated with a greater likelihood of a scarred heart.

Another study using laboratory rats provides some possible explanations for why. In the study, Canadian and Spanish scientists prodded young, healthy male rats to run at an intense pace, day after day, for three months, which is the equivalent of about ten years in human terms. The training was meant to mimic years of serious marathon training in people, says Stanley Nattel, M.D., a cardiologist at the Montreal Heart Institute Research Centre and one of the authors of the study.

The rats had started out with normal hearts. But after their long, hard marathon training, scans showed that most of the rodents had developed scarring in their hearts and some structural changes similar to those seen in human endurance athletes. The researchers could induce arrhythmias, or disruptions of the heart's natural electrical rhythm, much more readily in the running rats than in a control group of unexercised animals. Thankfully, when the animals stopped running, their hearts returned almost to normal within eight weeks. By and large, the fibrosis disappeared.

Should longtime marathoners worry, then, about the state of their hearts? At the moment, it's impossible to say. Too many factors remain unknown, including genetics, other health habits, and luck. "Let's say we ask a hundred people, all the same age, the same gender, to start a marathon training program at the age of twenty years," says Paul Volders, M.D., a cardiologist at Maastricht University, in the Netherlands, who's studied athletes' hearts. If the runners continue their training uninterrupted for thirty years and scientists then scan their hearts, he continues, "it is very likely and, in fact, one may say, it's a certainty that there will be major differences in the tissue of the chambers of the heart between these people." For some, the changes will be beneficial; for others, probably not.

If you are a lifelong marathon runner or other serious endurance athlete and your heart occasionally races, which could indicate arrhythmia, or in any other unusual ways draws attention to itself, Dr. Nattel says, consult a doctor.

But, realistically, most of us do not need to fret about exercise-induced cardiac fibrosis. "Most people just run to stay in shape, and for them, the evidence is quite strong that endurance exercise is good" for the heart, says Paul Thompson, M.D., the chief of cardiology at Hartford Hospital, in Connecticut, and an expert on sports cardiology. If you exercise regularly and currently have no heart symptoms, "I think it's safe to say that you should keep it up," he continues.

"How many people are going to join the 100 Marathon Club?" he asks, or undertake a comparable amount of training? "Not many," he says. "Too much exercise has not been a big problem in America."

How Much and What Kind of Exercise Do You Need? Some Hints

1. Take This Quiz.

It consists of a single question, courtesy of Dr. Frank Booth, one of the country's leading experts on exercise dosing. "Do you want to live to be a hundred?" Dr. Booth asks. Since we both know that you do, you've now established a rationale for using exercise to improve your health. "Being active is the best, easiest, and cheapest way to decrease all-cause mortality and increase functional life span," Dr. Booth says. "People who don't exercise are at greatly increased risk of dying earlier than they need to."

2. Got Health?

If enhanced physical and emotional well-being—i.e. health—and a centenarian's birthday party are your primary goals, then the current national exercise guidelines, issued in 2008 by the Department of Health and Human Services, apply to you. They suggest:

> **a.** 150 minutes (2 hours and 30 minutes) of moderate aerobic activity each week, such as brisk walking or lap swimming. Or

> **b.** 75 minutes (1 hour and 15 minutes) weekly of more vigorous aerobic activity, such as running. Plus

> **c.** Weight training at least twice a week, to ensure that all muscles are healthy.

You can partition the sessions in whatever way works for you. Walk for 30 minutes a day 5 times a week, or divide the walks into 15-minute blocks and go twice a day. The best science suggests that the body doesn't really care.

3. Want More?

If you have ambitions beyond glowing health; if, essentially, you are an athlete—which does not mean that you must compete, only that you ache to be a little faster or better at your chosen activity—you will have to push your body somewhat. You must overload the musculoskeletal and cardiovascular systems in order to improve your fitness and performance. You can do this with whatever type of exercise you already enjoy. Just do it harder than normal sometimes. An interesting study of walking conducted a few years ago divided a collection of about five hundred adults into multiple groups. All walked for thirty minutes at a time, but at differing intensities, based on their heart rates, or for different frequencies throughout the week. The group that walked at a pace equal to about 70 percent of their maximum heart rate, meaning quite briskly, had by far the greatest increase in physical fitness, compared with the groups that walked at lower intensities, even if they scheduled more walking sessions during the week. At the easy intensity, though, those who walked more often (six or seven sessions a week) wound up with better measures of aerobic fitness than those who walked four times a week. So to improve your physical fitness and performance:

a. Occasionally increase the intensity or frequency of your usual workouts.

b. Wear a heart rate monitor if you'd like, but multiple recent studies have shown that people are often better judges of their workout's difficulty than even the best gear. Maximum heart rates vary wildly, despite what those charts at health clubs suggest, and an exact calculation of yours would demand a treadmill test at an exercise physiology lab. Trust your intuition.

c. If you're breathing hard enough that you can barely converse during a workout, you're exercising vigorously.

4. Make It Snappy.
You can compress all of your exercise into a few minutes a week. This approach improves both health and fitness in a very

short time frame but is not for the fainthearted. (Which reminds me: If you haven't been exercising regularly until now, consult a doctor about your heart and other health issues before beginning, obviously.) HIIT or high-intensity interval training requires that all of your exercise, not just some, be intervals. Pioneered by Dr. Martin Gibala at McMaster University, in Canada, the original studies of HIIT involved grueling, abbreviated workouts conducted at a pace beyond each person's supposed maximal heart rate for thirty seconds at a time on specialized stationary bicycles known as Wingate ergometers. Most of us don't want to work that hard and have never heard of a Wingate. So recently Dr. Gibala tested a slightly more humane and practicable version of HIIT. The results were gratifying. When a group of young, untrained men practiced this new HIIT for two weeks, they significantly increased the molecular markers associated with aerobic fitness. Others have found that the HIIT protocol also lowers blood sugar and improves blood pressure control, meaning it's good for your health. To employ this greatest HIITs version, you need only a normal stationary bicycle, some sense of your maximum heart rate, and grit.

    a. Warm up with at least 3 minutes of easy cycling.

    b. Then HIIT it, with intervals consisting of 60 seconds of almost all-out pedaling, equivalent to almost 100 percent of your maximum heart rate. Basically make yourself as uncomfortable as you can stand for a minute. (For the cycling tech weenies out there, the average power wattage in Dr. Gibala's experiment was 355 watts during the sprints.)

    c. Rest with 75 seconds of low-intensity pedaling (below 40 percent of your maximum heart rate, or 30 watts).

d. Repeat the sprint-rest interval at least 8 times to start, working up to 12 or more repeats.

e. Pedal easily for a few minutes at the session's end, for a total time commitment during each workout of less than 30 minutes.

f. Aim to complete 3 sessions a week, or an hour and a half (or less) in total. "It would take at least five hours of conventional endurance exercise for the same benefits," Dr. Gibala says. "I personally don't have the time to go out and exercise for hours," he added. "That's one of the reasons we're working so hard" on perfecting HIIT.

5. Back Off.

There are times when either you can't keep up with your intended exercise or you know that you've been overdoing things. Muscular aches that don't go away after about three days, most exercise experts say, can indicate an incipient overuse injury. Slow down or stop exercising for the time being and consult a doctor or physical therapist. The good news is that reductions in exercise don't have to strip you of your hard-won health and fitness gains. In a recent study of collegiate female dancers, those who ceased all dance practices developed, within two months, noticeably higher blood sugar levels and (horrors) waist circumferences, but by and large, those who maintained a much reduced but continuing schedule of practice pliés did not. The same was true for average adults who'd completed four months of weight training and were asked, for a recent experiment, to reduce the number of their visits to the gym from three times weekly to once. Most maintained their strength gains for eight months. Few added additional strength or muscle mass, but they didn't lose everything either. A "once per week exercise dose was generally sufficient to maintain positive neuromuscular adaptations," the study authors concluded. Thank you, science.

6. Clean Up.

Finally, if improved mental health is a goal, cleaning the house seems, oddly, to help. In addition to the survey of Scotspeople in which housekeeping was one of the activities associated with less "psychological distress," another new, large-scale survey in Europe found "an inverse association between housework and distress." Vacuuming and mopping may make you happier. My house is available anytime for training.

2

Stretching the Truth

For a study published in late 2010, scientists at the Florida State University, in Tallahassee, recruited ten competitive male collegiate athletes and asked them not to stretch during their warm-ups. Many athletes would, of course, balk at that request. A separate, large-scale study of recreational runners, conducted under the auspices of the USA Track and Field Association and also published in 2010, required years to complete, because the researchers couldn't find a sufficiently large body of runners willing to give up their stretching routines, even in the interest of science. Thankfully, college-age men are, in general, not averse to avoiding physical exertion, even stretching, when they can, and so the ten men signed on.

The researchers brought the volunteers into the university's exercise physiology laboratory for a series of fitness tests, including measurements of their flexibility, then had them return for two additional sessions. During both, the men ran on a treadmill for an hour. In one session, they prepared for the run by simply sitting quietly for sixteen minutes. In the other, they stretched first, following a scripted sixteen-minute static-stretching routine. Static stretching (which is what most of us mean when we talk about stretching) involves stretching a muscle to its maximum length and holding it for twenty or thirty seconds. After stretching, the men felt more flexible.

But their performance declined, significantly. During the hour-long run, they covered less distance than when they had just sat quietly. They also consumed more calories and oxygen during the run, suggesting that their strides had become less economical, that the running was physiologically harder. The implication, the scientists concluded, was obvious. "Static stretching should be avoided before endurance events," they wrote. To which many athletes would respond, if their ongoing behavior is any indication, with an eye roll and a brush-off.

Gumby or Good

For a practical lesson in how long it takes for exercise science to penetrate to the playing fields, watch your child warm up for his or her next soccer match or spy on the trailing end of a marathon field, back where the less experienced runners bunch. The soccer players will almost certainly be warming up by bending over to touch their toes, at the urging of their coaches. The marathon runners will be steepling against light poles, statically stretching their hamstrings, or sitting on the pavement, levering their trunks toward their knees.

"It's pretty discouraging, really," says Duane Knudson, Ph.D., a professor of biomechanics at Texas State University who has been studying stretching and athletes for years.

Most of us learned how to warm up in grade school, by touching our toes and slowly stretching our muscles, and haven't changed our routines much since. Science, however, has moved on. In the past decade, a growing number of studies have shown that static stretching not only does not prepare muscles for activity; it almost certainly does the reverse. In a representative experiment conducted a few years ago at the University of Nevada, Las Vegas, athletes generated less force from their leg muscles after static stretching than they did without stretching. Other studies have found that stretching before exercise decreases strength in the stretched muscle by as much as 30 percent. Weirdly, stretching the muscles in one leg can even reduce strength in the other leg, an effect that can last for up to thirty minutes. In a few key real-world studies, basketball players who stretched before a game were unable to jump as high during play as when they hadn't stretched.

For runners and other endurance athletes, the utility of being limber at all is in question. In one recent, telling experiment, elite collegiate distance runners underwent tests of their hamstring flexibility. The runners, as a group, weren't especially supple, although this varied from person to person. Overall, the women were more limber. (They usually are.)

More surprising, when researchers compared the runners' flexibility scores to their best times in a 10K road race, those with the tightest, least flexible hamstrings tended to be the fastest. They also had the best running economy, meaning that they used the least energy to go the same distance as other runners. Probably, the researchers concluded, tighter leg muscles allow "for greater elastic energy storage and use" during each stride. Think of a rubber band. If it's overstretched and limp, it doesn't snap back when pulled and released. So, too, with your hamstrings: If they're loose, they don't efficiently lengthen, shorten, and snap back into place with each stride. To some degree, as an endurance athlete you can be as flexible as Gumby, or you can be good.

Of course, many of us might accept a temporary reduction in strength and even in speed if stretching protected us against injuries, as many of us have long believed that it must. But in multiple large-scale studies of military recruits during basic training, stretching before long marches and runs did not lessen the incidence of overuse injuries. In the largest of these studies, results showed that an almost equal number of soldiers developed lower-limb injuries (shin splints, stress fractures, etc.), regardless of whether they had performed static stretches before training sessions. Similarly, in the largest study to date of everyday athletes who stretched, almost fourteen hundred recreational runners aged thirteen to sixty-plus were assigned randomly to two groups. The first group did not stretch before their runs, while otherwise maintaining their normal workout and warm-up regimens. The second group did stretch.

Both groups followed their routines for three months. At the end of that time, quite a few of the runners had missed training days due to injury, a predictable result, since running is one of the most injury-plagued sports on the planet. But there was no difference in the final pain tally between the two groups. The same percentage of those who stretched injured themselves as those who didn't. Static stretching had been a wash in terms of protecting against injury, raising the obvious question: So why in the world do so many of us still warm up by stretching?

Flexibility Is Overrated

"It's been drummed into people that they must stretch, stretch, stretch—that they have to be flexible," Dr. Knudson says, and some subsets of people do require outsize degrees of flexibility, such as gymnasts, pole dancers, and politicians. For the rest of us, extreme suppleness is unlikely to bolster our performance and is probably unachievable, anyway. "To a large degree, flexibility is genetic," says Malachy McHugh, Ph.D., the director of research for the Nicholas Institute of Sports Medicine and Athletic Trauma, in New York, and an expert on flexibility. You're born stretchy or not. "Some small portion" of each person's flexibility "is adaptable," Dr. McHugh adds, "but it takes a long time and a lot of work to get even that small adaptation."

What, then, are we actually doing when we dutifully stretch before a workout? Both more and less, as it turns out, than we might wish. "You may feel as if you're able to stretch farther after holding a stretch for thirty seconds," Dr. McHugh says, "so you think you've increased that muscle's readiness." But typically you've increased only your mental tolerance for the discomfort of the stretch.

"There are two elements" involved in stretching a muscle, he continues. One is the muscle itself. The other is the mind, which sends various messages to the muscles and tendons telling them how to respond to your stretching. What changes as you stretch a muscle is primarily the neural message, not the physical structure of the muscle. The cells don't lengthen. Instead, you "start to develop a tolerance" for the discomfort of the stretch, Dr. McHugh says. Your brain will allow you to hold the pose longer. But only up to a point. As you continue to hold, the nervous system begins to fear that you'll tear the muscle and, in a protective reaction, causes a "neuro-muscular inhibitory response," he says. That response, meant to be a fail-safe, makes it harder for the muscle to contract with all of its force, rendering it weaker. (Some intriguing but not fully substantiated research has found that ballistic stretching, the old-fashioned, much-maligned style in which you bounce repeatedly through a stretch, actually prepares muscles better for sports performance than static stretching, because it doesn't ignite the neural inhibitors. But none of the recent studies of ballistic stretching have tested its long-term safety, so you still shouldn't try it at home, kids.)

Even the sensation of elasticity is short-lived, according to Dr. McHugh. In a review article of the effects of stretching, he looked at the measurable impacts of a number of different stretching regimens. What he found was that when people performed four ninety-second stretches of their hamstrings, their so-called passive resistance to the stretching decreased by about 18 percent—they felt looser—but the effect had passed in less than an hour and their muscles returned to the same level of flexibility as before. To achieve a longer-lasting impact and to stretch all of the muscles involved in running or other sports would probably require as much as an hour of concerted stretching every day, according to Dr. McHugh. "And the effects still wouldn't be permanent," he says. "You only see changes" in the actual physical structure of the muscle cells and surrounding matrix "after months of stretching, daily, for hours at a time. Most people aren't going to do that."

And most probably don't need to. "Flexibility is a functional thing," Dr. Knudson points out. "You only need enough range of motion in your joints to avoid injury. More is not necessarily better." The extremes of flexibility and inflexibility are where dangers lie. Most of us, though, are somewhere in the middle.

If you're worried about where you fit on the flexibility spectrum, try this simple test. Known accurately, if unimaginatively, as the "sit-and-reach test," it's a "pretty good" DIY evaluator of back and hamstring flexibility, Dr. Knudson says, and requires no fancy equipment. Simply sit at the bottom of a staircase. (If you don't have one at your house, substitute a heavy box.) Straighten your legs so that your feet push against the bottom step, toes upright. Stretch forward. "Try to lay your chest onto your thighs," he says. If you can reach to your toes, you're more than flexible enough. (No one yet has devised a way to reduce flexibility, by the way, although some Olympic-level coaches in other countries are rumored to be trying.)

If, on the other hand, "you can't get anywhere near your toes, and the lower part of your back is practically pointing backward" as you reach, then you might need to work a bit on your hamstring flexibility, Dr. Knudson says, to avoid injuring yourself while exercising or racing. Consult a physical therapist about proper stretches and technique. And be prepared for little overt progress. "You won't get a lot of change," Dr. Knudson says," but a little may be all you need."

Do You Need to Warm Up at All?

The fact that there are few benefits (and many disadvantages) to stretching before a workout does not necessarily mean that you shouldn't warm up before a workout. An impressively comprehensive recent analysis of more than forty years' worth of scientific literature about warm-ups found that, in aggregate, warming up was beneficial to performance across a wide range of sports. A well-designed warm-up improved athletic performance in seventy-nine different categories, from the speed with which sprinters covered one hundred meters to the power generated by rowers. Cyclists, distance runners, swimmers, and even bowlers in various studies improved their subsequent athletic performance if they warmed up. A separate study of golfers, a group notorious for not warming up at all, except perhaps by layering on an extra argyle sweater, produced similar results. When the golfers were randomized into groups that either statically stretched or warmed up (without stretching) by completing multiple practice swings, their games changed markedly. The golfers who had stretched produced puny drives out on the links, while those who had warmed up with practice swings drove the ball 7 percent farther, with 60 percent greater accuracy.

But there are problems with many of the studies that have looked at warm-ups, says Andrea Fradkin, Ph.D., a professor at Bloomsburg University, in Pennsylvania, and author of both the review and the golf study. Most have been small-scale and short-term, she says, and their methods inconsistent. In some, the volunteers stretched to warm up. In others, she says, "they did jumping jacks or passed a medicine ball around and then raced bicycles." None of the studies have proven that any one approach to warming up is best, or even that a warm-up invariably makes you better.

The science about how to warm up "is just not well advanced," Dr. Fradkin admits. "We haven't answered the big questions yet," about whether to warm up or why to warm up, "let alone the smaller, specific ones" about how.

Even the question of whether a thorough warm-up prevents injuries remains surprisingly open. A major study conducted several years ago by the Centers for Disease Control and Prevention found that knee injuries were cut nearly in half among female collegiate soccer players who followed a complicated warm-up program. But the accompanying balance-training program was probably the key to cutting injuries, not the warm-up. On the other hand, a smallish recent study of golfers showed that those who warmed up properly were nine times less likely to get hurt subsequently.

What, though, constitutes a proper warm-up? Most experts agree that a warm-up should do two things: increase the range of motion in the joints that will be used in the upcoming exercise, and literally warm up the body. When you're at rest, there's less blood flow to muscles and tendons, and they stiffen. "You need to make tissues and tendons compliant before beginning exercise," Dr. Knudson says. In a famous, if gruesome, study from a few years ago, scientists showed that leg-muscle tissue taken from laboratory rabbits and attached to tensioning machines could be stretched farther before ripping if it had been electronically stimulated—that is, moved and heated. It could withstand higher loads, similar to those that muscles experience during strenuous exercise, when the tissue was warm.

To raise the body's temperature, begin your warm-up with aerobic activity, usually light jogging. If you're planning only a brisk walk for your exercise session, you should still "warm up" with a few minutes of strolling, progressively boosting your pace.

Timing and intensity are considerations. A number of experiments suggest that competitive athletes often warm up too strenuously or too early. A 2002 study of collegiate volleyball players found that those who'd warmed up before a match and then sat on the bench for thirty minutes wound up with stiffer lower backs than before the warm-up. If you're not a starter, time your warm-up to coincide with the minutes before you'll enter the game.

Meanwhile, a number of recent studies have demonstrated that an overly vigorous aerobic warm-up simply makes you tired. Collegiate kayak racers, for instance, were slower after warming up with a brisk session on a rowing machine than when they warmed up with more languid machine rowing. Similarly, in a study from New Zealand, a group of untrained men in their twenties were brought into a human performance lab and allowed to determine the intensity of their warm-ups on stationary bicycles. They spontaneously started competing with one another and were subsequently too fatigued to complete a cycling power test, a result that may say more about twenty-year-old males than about the hazards of a too-strenuous warm-up.

More definitively, researchers at the University of Calgary, in Alberta, Canada, recently asked experienced bike racers to complete, on alternate days, their usual warm-up or a shorter, easier version. Cyclists are well known (even notorious) for the length and intricacy of their warm-ups. In this case, the volunteers, all highly trained racers, first followed their standard and very lengthy warm-up routine, beginning with twenty minutes of riding. The intensity of the pedaling increased until it reached about 95 percent of each rider's maximal heart rate. That session was followed immediately by four hard intervals, or timed sprints, during which the rider would pedal as hard as he could for eight minutes.

That warm-up represents more exertion than most of us willingly would expend in a full workout, and "we suspected that it might be harder than it should be," says Brian R. MacIntosh, Ph.D., a professor of kinesiology at the University of Calgary and master of understatement.

A year earlier, Dr. MacIntosh had studied sprint skaters' warm-ups as part of the national skating team's preparation for the 2010 Winter Olympics. The skaters "were warming up for two hours for a thirty-five-second race," Dr. MacIntosh says. Afterward, he found, the skaters' leg muscles contracted with less force than they'd generated before the warm-up. In warming up their muscles, they'd exhausted them.

The cyclists were doing the same. When the researchers stimulated the riders' leg muscles electrically, they found that the muscles contracted less forcefully after the warm-up than before.

A more leisurely, fifteen-minute warm-up left the riders' legs primed and fresh, as evidenced by their performance on a thirty-second all-out pedaling effort. They stomped the pedals with satisfying power, generating far more watts than after the prolonged warm-up. "This research provides an argument against the traditional 'more is better' warm-up concept that is adopted by many competitive athletes," Dr. MacIntosh wrote in the study, helpfully titled "Less Is More."

In general, the few reliable studies of warm-ups suggest that 10 to 15 minutes should be plenty, if you choose to warm up at all. Begin with an easy aerobic session paced at 40 to 50 percent of your maximum heart rate. Less technically, choose a pace at which you can sing. Progress to about 60 percent of your maximum heart rate, or a pace at which you can converse but not croon.

Then it's time to Spider-Man.

Exciting Your Muscles

Dynamic stretching may be the most important, if often neglected, element of a proper warm-up. Dynamic stretching, or, as many experts prefer to call it, dynamic movement or dynamic warm-up ("it's not stretching," Dr. Knudson says, "so calling it 'dynamic stretching' is a misnomer") involves moving and waking the tissues that will be called upon during the subsequent exercise. This allows you, in turn, to achieve adequate range of motion within your joints.

Elongating your muscles and tendons while you're moving does not seem to initiate the same kind of inhibitory neuromuscular messages that static stretching causes. Instead, dynamic stretching results in "excitatory messages" from nerve endings to the tissues, Dr. McHugh says.

Dynamic stretching is at its most effective when it's relatively sports specific. In other words, if you're going to run, you want, unsurprisingly, to prepare the muscles and connective tissues in the legs so that your knees, ankles, and hip joints are able to shift, twist, and absorb the forces generated by striking the pavement. If you're going to play tennis, you want to prepare not only your legs but also your shoulders, wrists, elbows, and back. In practice, this means that for runners an ideal warm-up might include linked squats, lunges, and form drills, such as kicking your buttocks with your heels. Athletes who need to move rapidly in different directions, such as soccer, tennis, or basketball players, should do dynamic stretches that involve many parts of the body. "Spider-Man" is a particularly good, if outlandish-looking, example: Drop onto all fours and crawl the width of the court or field, as if you were climbing a wall. This prepares you, too, for light cat burgling, should your athletic career flag. (There is a full dynamic stretching routine at the end of the chapter, as well.)

After whatever warm-up we do, our bodies should now be ready for a full workout, match, or race—after which we face the fraught issue of recovery, a topic marked by rooted but largely unsubstantiated thinking about the necessity of cooling down, the desirability of ice baths and deep massage, and the almost talismanic power of ibuprofen.

Dealing with DOMS

Exercise is, of course, physical stress and fairly Nietzschean at the molecular level. Muscles, bones, and connective tissues grow stronger by sustaining damage. Skeletal muscle in particular responds to unfamiliar exercise with a measure of harm. Skeletal muscle is a unique tissue, made up of long, thin fibers that are composed of several different proteins. (Cardiac muscle tissue is different, but that's not important here.) These proteins interlock like Legos inside fibrous compartments called sarcomeres. Sarcomeres can stretch, but only so far.

During certain kinds of movements, some of the sarcomeres within the affected muscles are pulled past their tolerance. The proteins inside separate, resulting, most physiologists believe, in micro-tears throughout your muscle tissue. Hours or even a day or two after the original exercise bout, this cellular-level damage is thought to lead to inflammation, the body's response to any invasion or insult to tissues, and strenuous exercise represents quite an insult. Blood vessels dilate in the affected sections of muscle, white blood cells and other immune-system-related cells flood in, tissues swell and warm, and, in general, mayhem reigns inside your muscles.

You almost certainly are familiar with the sensation. It's known as "delayed onset muscle soreness," or DOMS, and afflicts anyone who works out. Eccentric muscle contractions, during which forces are applied to muscles while they lengthen, as, for instance, when you lower a dumbbell, are the main culprit in delayed-onset muscle soreness. (A muscle can lengthen while it contracts, since *contraction* in this sense means creating force, not shortening.) Concentric contractions, in which muscles shorten—the upward motion of a biceps curl, for instance—cause less damage. That's why running downhill leaves you more sore the next day than running on flat ground.

In general, this soreness is a good thing. "You want to stress the muscles and connective tissues," says Thomas Swensen, Ph.D., a professor of exercise and sports science at Ithaca College in Ithaca, New York, and a leading researcher into exercise recovery. "They respond positively." The tissues rebuild themselves, becoming stronger and more pliable, a process known, rather blandly, as adaptation, which happens to be the foundation of fitness. "But," says Dr. Swensen, "it's only effective if you recover properly."

Which most of us probably do not.

A Downhill Battle

In a revealing study completed a few years ago, South African researchers asked volunteers to run for half an hour, backward, on treadmills that were set at a negative incline, as if the runners were striding downhill, a workout almost guaranteed to engender sore muscles. After thirty minutes of the exercise, half of the group was instructed to cool down by walking gently uphill on the treadmills. The other half stopped. They didn't formally cool down. They simply hopped off their treadmills and headed to the showers.

Exercise lore would assure us that those runners who didn't cool down would be much more sore the next day. Most of us have been told by coaches, training partners, or random strangers that a cooldown, usually with a diminished version of our full workout (a slow walk after a run, or a few minutes of easy pedaling at the end of a hard ride), is essential to keep muscles healthy. But in the South African study, there was no difference between the groups in their reported rates of sore muscles in subsequent days. Cooling down did not prevent DOMS (and, as a sub-lesson, running downhill backward really promotes it; almost all of the runners reported being quite sore).

"One of the main things that science can now tell us" about DOMS, Dr. Swensen says, "is that there's not much that works" to prevent or lessen the distress, and at least some of the solutions that we try may be counterproductive.

Take ibuprofen. You probably are. Countless people who exercise pop the nonsteroidal anti-inflammatory drugs (NSAIDs) regularly, hoping to avoid sore muscles. "For a lot of athletes, taking painkillers has become a ritual," says Stuart Warden, Ph.D., an associate professor of physical therapy at Indiana University, who has extensively studied the physiological impact of the drugs. "They put on their uniform" or pull on their running shoes and pop a few Advil. "It's like candy," or vitamin I, as some athletes refer to ibuprofen. In one recent study, seven out of ten runners at the Western States 100 ultra-endurance marathon took painkillers before the race, hoping to prevent sore legs afterward, while more than 60 percent of racers at a recent Brazilian Ironman reported that they planned to pop ibuprofen tablets before, during, and after the race. Marathoners also rely heavily on the pills. About 13 percent of participants in a 2002 marathon in New Zealand said they had swallowed NSAIDs prophylactically before the race, while the reported percentages have been even higher at other races, approaching 50 percent of participants in some cases. At the 2006 Men's World Cup soccer tournament, more than half of the elite players took ibuprofen or other NSAIDs at least once during the tournament, with more than 10 percent using them before every match.

The reason why ibuprofen is so popular is obvious. Athletes sincerely believe that the drugs will prevent sore muscles. The Brazilian Ironman triathletes, for example, almost uniformly cited "pain prevention" as their rationale for swallowing NSAIDs. Similarly, when the Western States runners were polled, most told the researchers that they thought ibuprofen would get them through the pain and discomfort of the race and stave off the development of searingly sore leg muscles afterward. But according to polling during and after the race, the runners using NSAIDs were just as sore during and after the event. Worse, they had higher levels of inflammatory markers than runners not using ibuprofen, even though the drugs should be anti-inflammatory. They showed signs, too, of mild kidney impairment and, nastily, low-level endotoxemia, a condition in which bacteria leak from the colon into the bloodstream.

The most disturbing news about ibuprofen, though, may be that the pills blunt exercise adaptations. In laboratory experiments on animal tissues, NSAIDs slowed the healing of injured muscles, tendons, ligaments, and bones. "NSAIDs work by inhibiting the production of prostaglandins," substances that are involved in pain and also in the creation of collagen, Dr. Warden says.

Collagen is the building block of most tissues. So fewer prostaglandins means less collagen, "which inhibits the healing of tissue and bone injuries," Dr. Warden says, including the micro-tears that accompany strenuous exercise. Essentially, "you lessen the training response" by taking painkillers. Your bones don't thicken or your tissues strengthen as much as they otherwise would. They may be less able to withstand the next workout. The pills that we take to reduce the chances that we'll feel sore may increase the odds that we'll wind up injured—and sore. In the end, there is "no rationale for the current prophylactic use of NSAIDs by athletes," Dr. Warden wrote in a recent editorial, "and such ritual use represents misuse."

But at least we can rely on ice baths and massage to effectively coddle our sore muscles. Right?

The Massage Myth

There are times when science makes mourners of us all. If you love a postexercise massage (and I did and do), the following experiment will be a blow. For the work, researchers in Canada recruited twelve healthy young men and asked them to exhaust their forearm muscles by continuously squeezing a specialized handgrip at 40 percent of their full strength for two minutes. "If that doesn't sound hard, try it," says Michael Tschakovsky, an associate professor in the School of Kinesiology and Health Studies, at Queen's University, in Kingston, Ontario, Canada, and the study's lead author. By the end of the two minutes, the men's arms shook with fatigue and their muscles were awash in lactic acid, a recognized by-product of strenuous exercise.

Most of us, including many physiologists, have long thought that the buildup of lactic acid contributes to postexercise muscle soreness, although, like so much else in exercise science, that theory is in doubt. Lactic acid, new research suggests, is at least in part a muscle fuel and so may be desirable in muscles after exercise, although not all scientists agree with that idea. For now, rightly or not, removing lactic acid remains one of the primary reasons that people schedule massages after exercise. The researchers wanted to see whether massaging the affected muscles would clear away the lactic acid more quickly than not massaging them. "We wanted to see if massage fulfills" its promise, Dr. Tschakovsky says.

After the volunteers had exhausted their arms and rendered their muscles full of lactic acid, they either had their arms massaged by a certified sports-massage therapist for ten minutes or lay quietly for the same ten minutes. Throughout, blood flow to their forearm muscles was monitored, while their lactic acid concentrations were checked via blood samples.

To almost everyone's surprise, particularly the athletes', massage did not increase blood flow to the tired muscles; it reduced it. Every stroke bore down on large and small blood vessels in the muscles, cutting off flow. Although the stream returned to normal between strokes, the net effect was to significantly lessen the amount of blood that reached the muscle compared with blood flow in the non-massaged group. Massage "actually impairs removal of lactic acid from exercised muscle," Dr. Tschakovsky and his colleagues concluded—a finding that, sadly, is in agreement with other new research about sports massage. A number of other studies have found that massage impedes rather than improves blood flow to tired muscles, and that kneading muscles does not speed their recovery. A typical experiment on boxers found that when they were provided with massages between matches, they reported feeling more relaxed. But they didn't perform any better in subsequent bouts than when they hadn't been massaged. The stroking and kneading had felt quite nice, but it had conferred no physiological benefits.

None of which means that massages are worthless. "Our study was only designed to see whether a massage will remove lactic acid from tired muscles, which is generally the reason that athletes, at least, want a massage," Dr. Tschakovsky says. "It doesn't work for that purpose. But that doesn't mean a massage has no benefits. It just means that we don't know yet what they are."

No Ice

Meanwhile, you may not be astonished by this time to learn that ice baths, another popular recovery method, also have, at best, questionable utility. "Taking a post-exercise plunge into an ice bath appears to be a common practice among many elite athletes," a recent study of the practice reported. The frigid water is thought to lessen postexercise swelling and in various other ways reduce soreness and encourage recovery. But in one of the few randomized controlled trials of the practice, people who sat in an ice bath after hopping on one leg to exhaustion had just as much pain and swelling the next day in their pogo-ing leg as a separate control group that had sat quietly after hopping. In fact, the ice bathers reported more pain than the control group during a test in which they rose out of a chair using their tired leg for support. "The protocol of ice-water immersion" in the study "was ineffectual in minimizing markers of DOMS," the authors concluded.

Similarly, in a recent case-study article, a rather bemused-

sounding emergency room physician reported treating several athletes—one a martial arts fighter, the other a marathoner—who'd
worked out strenuously, then dutifully soaked in ice, and wound up
so sore the following day that they had felt obliged to visit an urgent
care facility. "This practice" of immersing oneself in ice after exercise
"may *cause* muscle soreness the day after," the doctor wrote (italics
added).

None of which seems to be deterring serious athletes, who, with
dauntless optimism, are taking ice baths to a new and otherworldly
level, thanks to whole-body cryotherapy chambers. Originally intended to treat certain medical conditions, whole-body cryotherapy
chambers are sealed rooms where the ambient temperature is lowered to a numbing −110 Celsius, or −166 Fahrenheit. They are essentially walk-in freezers. Many elite athletes have begun using the
frigid chambers in hopes that supra-subzero temperatures will help
them to recover from strenuous workouts more rapidly.

Before entering a cryochamber, users must strip to shorts or a
bathing suit, remove all jewelry, and don several pairs of gloves, a
face mask, a woolly headband, and dry socks. (Justin Gatlin, an
American sprinter, neglected that last precaution while using a cryochamber in the weeks before the 2011 track and field World Championships; his socks were sweaty from a previous workout and froze
instantly to his feet. He arrived at the championships with frostbite
and did not make the finals of the 100-meter dash.) The athletes
then move through an acclimatization chamber set to about −76
Fahrenheit and from there into the surface-of-the-moon-chilly cryotherapy chamber.

At −110 degrees Celsius, whole-body cryotherapy is "colder
than any temperature ever experienced or recorded on earth," says
Joseph Costello, Ph.D., a researcher at the University of Limerick,
in Ireland, who is studying its effects. The athletes remain in the
chamber for no more than two or three minutes, stamping their feet
and waving their arms to retain circulation. A Welsh rugby player

described the experience as being in an "evil" sauna, but told British reporters that he believed that the sessions were helping him to recover more quickly from rigorous practices.

But perhaps addled by the cold, he may have been deluding himself. A study by Dr. Costello found that whole-body cryotherapy did not lessen muscle damage among a group of volunteers who'd completed grueling resistance exercises with their legs before entering the chamber. They recovered no quicker than a control group who did not expose themselves to the coldest temperatures on earth.

Still, there is something about icing sore muscles that has undeniable appeal. In a recent study, young men who completed a punishing ninety-minute shuttle run and then eased themselves into a frigid bathtub (with the water cooled to fifty degrees Fahrenheit) for ten minutes wound up with the same levels of creatine kinase, a hallmark of muscle damage, as runners who didn't soak. But they said that they felt less sore than the others. So, too, a group of competitive cyclists reported that their legs definitely "felt better" when they had an ice bath after a strenuous cycling time trial than when they didn't, even though, in harsh reality, they performed no better on a second time trial after sitting in an ice-filled tub than after recovering by simply sitting.

And that result, in its ambiguity, may encapsulate much of what you need to know about short-term recovery on a practical level. It's true that almost nothing is proven to speed healing or reduce soreness right after a hard workout. But many techniques, from massages to ice soaks, feel nice, and such hedonistic effects can't be dismissed. They "are not nothing," Dr. Swensen says. If scheduling a post-run massage makes the run itself seem more palatable, then by all means, call a masseuse. For some people, it may even be worse to change their routine than to continue it, no matter what the science says. When a group of dedicated competitive runners were asked recently not to stretch after their runs, for instance, they reported a disproportionately high rate of subsequent injuries. Few of the injuries were verified during physical exams; many probably were psychosomatic. But the runners were convinced that post-run stretching had kept them injury-free, and when they couldn't do it, they convinced themselves that they were getting hurt. Magical thinking can be potent.

Join the Circus

There is, however, one intervention that has been shown, with some reliability, to aid in recovery from vigorous exercise: refraining from engaging in vigorous exercise for a while—i.e. resting. Fitness develops, remember, through adaptation. You stress the system, it responds and becomes stronger. But if the same stress is imposed over and over and the system gets no opportunity to respond and strengthen, your body doesn't become stronger. The various microtraumas accumulate and become macro-traumas. The body breaks down.

So, rest. The question is how much. The answer, it would seem, depends on what you're doing in your workouts at the moment and what you ultimately hope to accomplish. "If an exercise session isn't unusually intense or unaccustomed," Dr. Swensen says, your muscles should already be used to it. They will have adapted to that amount of stress and you aren't likely to develop much consequent soreness. You therefore don't need much recuperative rest. Of course, avoiding unaccustomed amounts or intensities of exercise also means that you won't progress, in terms of fitness. You will have plateaued. That may be fine. If you jog three or four easy miles a day, don't dream of marathons, don't feel a burning need to be swifter, and scrupulously avoid hills, you and I are soul mates. More important, "you probably don't need to think much about formal rest days," Dr. Swensen says. A day a week of light-duty exercise, such as a walk instead of a jog, should allow any slight muscular damage to heal easily.

If, though, you exercise ambitiously and progressively, following a program of steadily increasing mileage or intensity, "you must rest regularly," Dr. Swensen says. But perhaps the most rigorous recent study of how much rest is best produced surprising results. The study was conducted on a group of proto-athletes, Cirque du Soleil performers, who frequently combine, in one package, strength, endurance, and otherworldly elasticity. They also follow a conveniently regular schedule of rest days, making them a useful study population. According to the study, a typical tour schedule includes one day off a week in the early weeks of a tour, followed by a two-day break, then more performances interspersed with single days off each week, and eventually another two-day break. The two-day breaks "are like mini-vacations and were instituted to alleviate physical and mental fatigue with the objective of protecting the artist from increased injuries and illnesses," the authors wrote.

But contrary to expectations, when the scientists examined incidence reports filed by the performers, they found that injuries were more prevalent in the days immediately after a two-day break, compared with after single rest days, among both strength performers (who snatch a partner out of the air on the trapeze) and those whose contributions were more endurance and agility based (acrobats and clowns). The authors reflected that this counterintuitive finding seems to show that two days off in a row may adversely affect "timing" and "precision." Ultimately, "one day per week rest periods for four to six week periods may be sufficient to prevent injury in subjects subjected to highly acrobatic and athletic physical demands."

The Cirque researchers aren't alone in concluding that too much rest may be unnecessary or undesirable. In one of the largest studies to date of the usefulness of rest days, military recruits who abstained from running for a week early in their basic training were more likely to become injured than runners who continued training with much shorter rest periods.

The message of the studies seems to be that rest should be deployed strategically. Lounge about for too many days in a row and you risk losing some of the hard-won neuromuscular adaptations that your body developed in response to earlier exercise. Then when you return to the same or a heightened level of exercise, you might have primed your muscles for injury.

Use common sense, as well as a dollop of science, to decide when and whether you've rested enough. "If you still feel twinges of soreness in your legs after one day of rest or have a lingering sense of fatigue, rest more," Dr. Swensen says. "It's not an exact science. Be aware of how you feel." Once the soreness abates, you should be ready to exercise again. As the authors of the Cirque study note, "a balance between training and rest is necessary to prevent the negative consequences of high-level activity." Monitor your muscles and they'll tell you when it's time to rejoin the circus.

You're Getting Warmer: Suggestions for a Proper Warm-up

1. Begin with Five to Ten Minutes of Easy Aerobic Exercise.

Jog if you'll be running; walk slowly if you're planning a brisk walk. Most sports players will probably find it most convenient to jog around the court or field, but bicycling there would work, too. The point is to elevate your heart rate and breathing gradually, so that your core body temperature rises gently. Keep the pace slow. There's no point in tiring yourself before you really begin exercising.

2. Segue Into Dynamic Stretching

At this point, you want to awaken and activate the various tissues that attach to your joints. Ideally, you should complete four or five dynamic stretches, concentrating on the joints you'll be using in your particular sport. The moves don't have to be difficult or unfamiliar. Jumping jacks, for instance, activate the ankle, knee, and, to some degree, the shoulder joints. Start with a few of those. Then incorporate some of the following drills, which represent dynamic stretches appropriate for a variety of sports. Adopt whichever meet your particular needs, and have fun with them. They may be called "dynamic stretches" now, but they're really just playground moves, legitimized by science. Some examples:

- **Skipping.** You know this drill. Raise your knee and skip forward. Try to land flat-footed (not on your toes) and push into the ground with each step. Do the same with your other leg. Repeat this sequence ten times or so. This is good for preparing the ankle, knee, and hip joints for running, cycling, and similar exercise.

- **Backward Skipping.** The same movement, in reverse. Check for obstacles behind you, of course, particularly fellow exercisers.

- **Bottom Kicks.** Rapidly raise one heel at a time toward your hindquarters, making sure your knees, thighs, and shoulders are in a straight line. Start slowly and increase the tempo of your kicks, priming your nervous system, as well as your muscles and joints for activities such as running, soccer, basketball, and tennis. Repeat 10 times with each leg.

- **Straight-Leg March,** aka the Little Toy Soldier. Kick one leg straight in front of you, with your toes flexed toward the sky. Reach your opposite arm to the upturned toes. Repeat with the other leg and arm.

Complete the full sequence about 10 times or until you cross the field or court where you will be playing.

• **The Scorpion.** So named because you thrash from side to side like a scorpion's tail. Lie on your stomach, with your arms outstretched and your ankles flexed so that your toes touch the ground. Kick your right foot over your back toward your left side, as far as is comfortable. Then switch and kick your left foot over your back toward your right side. Do the exercise only once at first, then work up to ten to twelve quick repetitions. An excellent way to warm up your lower back and hip joint, it's also a difficult, advanced exercise. If after a few attempts it feels uncomfortable or painful, skip it. Try the milder Mud Wallow instead: Lie on your back, draw your knees to your chest, and roll gently from side to side.

• **Handwalks.** An especially good drill for tennis and basketball players, handwalks warm up your shoulders, back, and legs. Stand straight, with your legs together. Bend over until both hands are flat on the ground. "Walk" your hands forward until your back is almost extended. Keeping your legs straight, inch your feet toward your hands, then walk your hands forward again. Repeat eight to ten times.

• **Practice Swings.** An astonishingly simple dynamic warm-up for golfers (many of whom skip even this rudimentary practice), according to a recent study, which found that this type of warm-up could improve driving accuracy by as much as 60 percent. Move your arms, shoulders, and back through the proper swing motion, without holding any weight. Then progress "through 'the bag,'" the study's authors wrote, "from shorter, heavier clubs to longer, lighter clubs, eventually reaching playing speed and intensity."

3. Be Judicious.

A warm-up should be a prelude to exertion, not the exertion itself. In Dr. MacIntosh's memorable study of competitive cyclists, those who completed a long, intense warm-up performed more poorly during a subsequent cycling trial than if they hadn't warmed up at all. How do you determine if your warm-up is too intense? If, like mine, yours consists of walking out the front door and then walking some more until you can convince your reluctant legs to break into a measured jog, you presumably are within safe parameters. Otherwise, "I wish I could tell you, one way or the other," Dr. MacIntosh, says. "But we don't know. My guess is that, if you've been following a warm-up routine for a while and you haven't gotten injured and you perform well when you race, then that warm-up is working for you. Keep it up." On the other hand, if your legs feel heavy and slow after your warm-up, you might be overdoing it. Experiment with shortening and reducing the intensity of your warm-up, Dr. MacIntosh suggests. "And if you find something that works, let us know," he urged. You will notably have advanced the state of knowledge in this particular field of exercise science.

4. Skip a Warm-up, If You Wish.

Should you choose not to warm up at all, you will be in fine company. Jack LaLanne famously scoffed at the idea of pre-exercising. "Warming up is the biggest bunch of horseshit I've ever heard in my life," he once told an interviewer. "Fifteen minutes to warm up! Does a lion warm up when he's hungry? 'Uh-oh, here comes an antelope. Better warm up.' No! He just goes out and eats the sucker."

3

It's Not About the Bites

During a brief period of my young adulthood, I took up bike racing. Back in those innocent days, fifty-mile training rides seemed fun. They also gave me the chance to trail behind and learn from more experienced racers. This was before the advent of portable energy bars and energy goo, so riders carried actual, movable feasts: peanut butter sandwiches, bananas, chocolate chip cookies, cold pizza, Pop-Tarts, slices of pecan pie, trail mix, and, on many a morning ride, Styrofoam cups of rapidly cooling joe. All were unwrapped, chewed, or sipped underway, sans hands on handlebars, a skill I still sometimes employ to impress my son. On hot days, some riders would slip unwrapped Popsicles into their water bottles, then slurp at the slurry as the temperatures rose, or they would carry sticks of frozen Pedialyte in their jersey pockets, letting it liquefy as the miles slid past.

I didn't take to road bike racing—too much speed, too many potholes, too little body armor, and my own well-deserved reputation for gaucherie. (Once, when my son was quite young and in a morbid stage, he ruminated on what would happen should his parents pass away. "If Dad died," he said, considering, "I'd only have you, Mom, and you are kind of clumsy.") But the experience did teach me about the value of food, real food, for athletes. In the intervening years, exercise science and many coaches and athletes moved away from that idea. Quite a few are heading back now. Sports nutrition is, in fact, in the midst of an interesting shift in perspective, with new science suggesting that sugar isn't bad for athletes, while too much water is; chocolate milk is an ideal supplement, and antioxidants are not; and good old familiar carbo-loading makes you fat. The interplay of nutrition and exercise is turning out to be fundamentally more complicated, intricate, and counterintuitive—while at the same time commonsensical—than most of us would have dreamed back in the days when Pop-Tarts in a jersey pocket were felt to be advanced nutritional planning.

The Wisdom of the Body

One of the classic texts in physiology is Professor Walter Cannon's 1932 book, *The Wisdom of the Body*, in which he popularized the idea of homeostasis, or the desire of the body to keep itself in balance and steady its operations. "Somehow," he wrote, "the unstable stuff of which we are composed has learned the trick of maintaining stability." This is a lesson that many of us profitably could remember when we start messing about with nutrition and exercise.

"There's always a new fad in sports nutrition," says Nancy Clark, a registered dietitian and certified sports nutritionist, who has long been one of the voices of quiet reason in the field. "Most turn out not to have lasting value." They may get raves from early athletic adopters. There may be enthusiastic, anecdotal endorsements. There may even be successes; some people will, at first, perform better on even the most outré of dietary regimens, "if only because they're expecting to," Clark says. "But most of these fads don't stand up to scientific scrutiny."

The basic mechanisms involved in sports nutrition are simple enough. During exercise, you move. Your muscles contract, burning energy (i.e. calories) and producing heat, which must be dissipated, in large part through sweat. So you lose fluids. The fluids and calories must be replenished (unless, of course, you're aiming for negative energy balance or weight loss).

It's in the fine details that controversy arises. What kinds of calories provide the best fuel: carbohydrates, fats, proteins? When should you ingest them? How much fluid do you need? What about sports drinks and electrolytes? What *are* electrolytes? And if a nutritional strategy works for most people—particularly if it works for most men—does that mean that it will work for me?

A Load of Carbs

Not long ago, physiologists at the Australian Institute of Sport and several other institutions systematically tested the concept of carbo-loading. Carbo-loading, which most of us have heard of and many of us have tried, gained traction back in the 1970s after studies showed that if you drastically cut back on your carbohydrate intake five or six days before a competitive event, then in the following days ate mostly carbohydrates (gigantic servings of pasta) while reducing your exercise volume (sitting around instead), you'd pack your muscles with carbohydrates, in the form of glycogen, muscles' preferred fuel, and presumably be able to run or cycle farther than you would without the extra fuel. Carbo-loading still is quite popular. Anyone who's run a marathon has probably been invited to a ceremonial carbo-loading pasta dinner the night before.

But when the Australian researchers assigned trained male cyclists to either a carbohydrate-rich diet or a placebo version—one that was, essentially, the riders' usual diet, but with the addition of a sweet shake the riders thought was filled with sugary carbohydrates but that actually was sugar-free—the results were a surprise. The riders eating the high-carbohydrate diet wound up with more carbohydrates stored in their muscles than the placebo group. But those extra carbohydrate stores didn't make them better riders. There were no statistical differences in the two groups' performances on a series of subsequent time trials, as long as both groups had access to sports drinks with carbohydrates in them. In that case, the cyclists' bodies preferentially burned the calories from the sports drinks, leaving any stored carbohydrates untouched. The carbo-loading had been unnecessary.

But it had almost certainly left the carbo-loaded riders fat. Stored carbohydrates pull water into muscle cells. The resulting weight gain may be temporary, mostly a few pounds of water weight, but it'll be present during whatever event you're carbo-loading for. Those enormous servings of pasta can't ensure that you'll be a better marathoner, but they can make you a heftier one. There is no such thing as a free lunch in sports nutrition.

Carbo-loading also has a decided gender bias. The studies I've mentioned used men as subjects; apparently, you'll find more of them hanging around physiological labs. But when researchers recruited female athletes, for a famous series of studies of sports nutrition, they found that women did not pack carbohydrates into their muscles as men did. Even when the women upped their total calories as well as the percentage of their diet devoted to carbohydrates, they loaded only about half as much extra fuel into their muscles as the men did.

Still, carbohydrates are important for athletes and, to some degree, casual exercisers. For fuel, contracting muscles rely on adenosine triphosphate, or ATP, a specialized molecule that provides energy. ATP is created by mitochondria, the microscopic powerhouses of your cells. Mitochondria find carbohydrates easy to work with, once the carbohydrates have been broken into simple sugars. When you start to move a muscle, the mitochondria within the muscle cells swing into action and use either the sugars in your muscles (technically known as glycogen) or sugars circulating in the bloodstream to make ATP.

Mitochondria can also make ATP from fat, either the lumpy stuff stored throughout your body or fats in the meal you just ate, but the process is more complicated and so the body usually prefers to burn through any available carbohydrates first.

Which does not mean you need to stash emergency candy bars in your pockets to get you through your next brisk walk.

Carb-Carrying Athletes

The reality is that if you're exercising for an hour or less, "the amount of additional carbohydrates you need is minimal," says Asker Jeu-kendrup, Ph.D., the global senior director of the Gatorade Sport Science Institute and one of the world's leading experts on sports nutrition, as well as an accomplished triathlete. If you're not a competitive or serious athlete, you don't need to be concerned with sports nutrition. A normal, healthy diet will supply more than enough carbohydrates to fuel moderate exercise lasting less than ninety minutes or so. A morning muffin will fuel you through a two-mile jog, easily.

If, on the other hand, you'll be working out for two hours or more—i.e. you've got a marathon in your future—you'll need more fuel. When to start that fueling process is somewhat controversial, though. For decades, most of us were told not to eat carbohydrates an hour before exercising, in part to sidestep unpleasant gastrointestinal consequences but even more to avoid the risk of "rebound" low-blood sugar, and the attendant wobbles and fatigue.

This concern grew out of decades-old studies showing that blood-sugar levels did decline if athletes swallowed sugary foods or drinks just prior to exercise.

But more recent and better-designed experiments found that while rebound hypoglycemia can occur, it's rare and doesn't affect performance much. When a group of British cyclists gulped sugary drinks before a workout, a few developed clinically low blood sugar in the first ten minutes of the bout, but the condition resolved soon afterward and they rode as well in a timed race as the other cyclists. Additional studies have found that eating easily digestible carbohydrates such as bananas or (a personal favorite) a Popsicle in the hour before exercise should enable you to work out longer. So don't bother starving yourself before your next workout.

But in general, the carbohydrates that you eat *during* a long workout or race are the most important for performance. As a rule of thumb, "you will probably want to ingest about sixty grams of carbohydrates per hour" during each workout and during the event itself, Dr. Jeukendrup says. "And when you engage in endurance exercise of two and a half hours or longer," he adds, "I would recommend a higher intake, up to ninety grams per hour."

That's a daunting task, by the way. It might sound delightful to be told to eat lots and lots of high-carb foods, but in practice, his advice translates into swallowing about a half gallon of a typical sports drink every hour while you're running or riding. Good luck with that. Easier, probably, is to rely on those totable sports gels and bars, if you can stomach them. In a series of elegant recent experiments, Dr. Jeukendrup and his colleagues showed that when trained cyclists rode at a moderate pace for three hours, they metabolized carbohydrates identically, whether they came from gels, sports bars, or sports drinks. But the bars and gels, in sufficient amounts, were much easier to carry.

They also conferred a competitive advantage, albeit with side effects. In a study of Ironman racers that bordered on the indecorous, those racers who consumed the most carbohydrates during events consistently placed higher than those who swallowed less. But they also had "increased scores for nausea and flatulence," the study's authors reported. What athletes will do to win (and what intrusive scientists will ask about).

Is Fructose Evil?

Interestingly, in the most comprehensive studies to date of carbohydrates and performance, athletes benefited most when their drinks, bars, or gels were sweetened not only with glucose (the simplest sugar, usually maltodextrin) but also with fructose, often in the form of high-fructose corn syrup. The two sweeteners together provided "the most energy," Dr. Jeukendrup says, presumably because they activate different metabolic pathways within the gut.

Fructose, though, has earned a decidedly unsavory reputation in recent years. Media coverage has suggested, with good reason, that our national overindulgence in fructose and other sugars is driving the epidemic of obesity, diabetes, and other illnesses. But that phenomenon largely involves people who are inactive. Sugars, and in particular fructose, behave differently in a body in motion.

Consider the results of a recent study of highly trained cyclists and their livers. For the experiment, Swiss and British researchers directed the cyclists, all men, to ride to exhaustion on several different occasions. After each ride, the men swallowed drinks sweetened with fructose or glucose. (Some also drank a milk-sugar sweetener.)

The liver is often overlooked when we contemplate organs integral to exercise, but it is an important reservoir of glycogen, or stored glucose. All sugars, including sucrose (table sugar) and fructose, are converted into glucose and stored as glycogen in the body. Strenuous exercise diminishes or exhausts the liver's glycogen, and until those stores are replenished, the body isn't fully ready for another hard exercise bout.

In this study, the scientists measured the size of each rider's liver (using MRIs) before and after the rides. All of the cyclists lost liver volume during their workouts, a sign that their glycogen stores were gone. But those who afterward drank fructose replaced the lost volume rapidly, regaining 9 percent after six and a half hours versus a 2 percent gain among riders drinking glucose-sweetened drinks. Overall, the researchers concluded, fructose-sweetened drinks were twice as effective in stimulating the liver to recover.

In general, fructose seems desirable for athletes. When cyclists in one large study downed a sports drink sweetened only with glucose during a two-hour bout of moderate pedaling, they rode faster during a subsequent time trial than riders who had drunk only water. But if the sports drink contained both glucose and fructose (in a two-to-one ratio), the riders were 8 percent faster in the time trial than those drinking glucose-sweetened fluids alone. (Most bottled sports drinks on the American market are sweetened with high fructose corn syrup, so they contain glucose and fructose in an almost one-to-one ratio.)

So sugars help you to exercise better. At the same time, and since nature is fair about these things, exercise also helps your body to deal with any health impacts from ingesting so much sweetness. Activity can "significantly reduce the health risks associated with fructose and other forms of sugar," says Richard J. Johnson, M.D., a professor of medicine at the University of Colorado, Denver, who has long studied fructose metabolism.

Let's look again at the liver. In sedentary people, ingesting large amounts of fructose, which is mostly metabolized in the liver, has been associated with the development of a disorder known as fatty liver. That condition can reduce the body's ability to respond to insulin, the hormone that helps to control blood sugar. A person with a fatty liver often develops resistance to insulin, becomes less able to control levels of glucose in the blood, and drifts almost inexorably toward type 2 diabetes.

But exercise can protect your liver from larding. Multiple studies have shown that even moderate exercise dramatically lowers the risk of developing a fatty liver, even if someone is overweight. A review of recent studies concluded that beginning an exercise program could significantly lessen the amount of fat in someone's liver, even if that person didn't lose weight during the program.

Overall, Dr. Johnson says, the "current science suggests that exercise exerts a positive physiological influence" on some of the same metabolic pathways that sugar harms. "Exercise may make you resistant to the undesirable effects of sugar," he says.

Not that any of us should live on sweets. "Sugar is not all bad," Dr. Johnson concluded, "but let's not pretend that it's healthy. It has a role" to play in prolonged activity, "but that role is limited. Sugar should not be a mainstay in anyone's diet."

Fructose and other sweeteners, then, are a bugaboo in sports nutrition. If you're the type of person who regularly rides a bike for three hours or more, fructose is your friend.

For anyone else, it's a lot of sugar.

Such contradictions are enough to drive anyone to drink.

A Thirst for Knowledge

In fact, the issue of how, when, and what to drink if you're exercising may be even more confusing and riddled with myth than the question of how and what to eat. "Fluid replacement seemed like a simple issue once," says Edward Coyle, Ph.D., an exercise physiologist at the University of Texas at Austin who's studied the topic for decades and served for many years as a consultant to cyclist Lance Armstrong. "It doesn't seem simple anymore."

To underscore the complexities, consider the findings of a recent study of the drinking habits of Kenyan runners. For years, as you no doubt know, Kenyans have dominated distance running. At one point, more than half of the top ten performances worldwide in distance races were run by Kenyans; this from a land with barely half a percent of the world's population. Scientists, coaches, and other athletes long have wondered what their training secrets are (and how to poach them). This study was eye-opening in that regard.

For the experiment, researchers from the International Centre for East African Running Science, in Scotland, visited a high-altitude training camp in Kenya during the summer. The campers in attendance consisted of many of the nation's elites, including Olympic and world champions. For five days, the researchers weighed each runner every few hours, while monitoring his or her urine output and its composition. Elite athletes get used to public scrutiny of their bodily functions; it comes with the territory. During workouts, the runners wore patches that measured how much they were sweating. They also kept scrupulous diaries of what they ate and drank throughout the day.

When the researchers compared the amount of fluid that the runners were losing through sweat each workout with how much they drank during that same run, the numbers didn't match. The runners sweated a lot but drank almost nothing, a practice that runs counter to current wisdom and practice among most of us here in America. The Kenyans typically lost several pounds during each workout. But in the evenings they'd swill cup after cup of weak bush tea fortified with milk and return refreshed and back to their prior weight the next morning. Not long after the study ended, several of the runners decimated the field at the track and field World Championships.

The apparent moral of the study? Weak, milky bush tea is the beverage of champions! And you obviously shouldn't drink anything at all during exercise.

The truer message, though, is that science is still untangling the realities from the folklore about an active body's needs for fluids.

Do We Need to Drink?

As you no doubt remember from grade school, about 55 to 60 percent of the human body is composed of water. Lean muscle tissue is nearly three-quarters water, and even solid-seeming bone contains more than 20 percent liquid. The stuff is essential for movement. Under various conditions, bodily water absorbs shocks around joints and organs, lubricates body parts, carries nutrients and waste products, and helps to regulate body heat. More fundamentally, scientists in 2010 announced that water molecules seem to be the agent (speculated about but never, until then, identified) that jump-starts the biochemical processes that end with the release of energy from ATP. Without water stores, the body can't produce the energy needed to propel it to a drinking fountain.

But for many years, athletes were advised to avoid water or other fluids during activity, no matter how long a session lasted. A racer who won the fifty-five-mile Comrades Ultramarathon, in South Africa, in the 1950s once told researchers that running "a complete marathon without any fluid replacement is regarded as the ultimate aim of most runners, and a test of their fitness." In the early days of modern marathon racing, water stations along the course were uncommon and widely spaced, and until the 1970s, riders in the Tour de France tended to drain only a few water bottles during five- or six-hour-long stages.

Attitudes about how much and when exercisers should drink began to change in the late 1970s, when the first sports drink, Gatorade, arrived in stores. While drinking an early version of Gatorade between plays, the University of Florida Gators had become national football champions. So it was easy to market this new beverage as a proven winner, a means for athletes to easily improve performance. At the same time (and in part thanks to funding from Gatorade), a number of studies were published that linked dehydration to potential heat illness in athletes. In the experiments, athletes who didn't drink fluids during exercise wound up with higher core body temperatures than those who did.

Soon athletes were being told by scientists and coaches that they should drink constantly during exercise sessions, imbibing enough to replace all fluids lost to sweat. The American College of Sports Medicine released guidelines suggesting that athletes should "drink as much as can be tolerated." Marathons added water and aid stations, often situating them every mile along the course instead of every two or three miles, as previously.

All of which seemed like a fine idea, until Oprah took up marathon running.

A Drinking Problem

Oprah Winfrey's famous 1994 marathon run inspired a tidal wave of imitators, many of whom had never attempted anything remotely so athletic as running 26.2 miles. Prior to Oprah's marathon debut, the average female marathoner finished the race in a little more than four hours. Within a few years after Oprah's marathon, the average finishing time nationwide for female marathon racers had risen to more than five hours, and it's continued to go up since. Average finishing times for men also have soared. Marathoners, as a group, have gotten much slower. Many of them don't "race" the course anymore, as the leaders do. They jog, walk, chat, make a day of it, and drink at every aid station along the way.

Then, at the 1998 Chicago Marathon, a forty-three-year-old pediatric dentist and mother of three died.

"This is that rare and strange instance where experts created a condition that had never existed before," says Timothy Noakes, Ph.D., a professor of exercise and sports science at the University of Cape Town in South Africa and a seminal if divisive figure in hydration science. "Thanks to the experts, people were warned that becoming dehydrated during exercise would harm them, but the bigger problem turned out to be drinking too much during exercise."

The Chicago marathon runner had died of hyponatremia, or water intoxication, a dangerous condition caused when someone drinks more fluids than he or she is losing to sweat. It's not hard to over-drink during a marathon if you're a slowish runner. Someone who jogs and walks at a leisurely pace for five or six hours along the course doesn't build up a lot of internal heat and doesn't, consequently, sweat much. But he or she is likely to drink at every aid station, and probably carries a water bottle, too.

Over several hours, over-drinking of fluids can dilute the bloodstream, lowering the concentration of sodium. (Think of a glass half-full of salt water. Add plain water to it and the contents become less salty; the sodium concentration drops.) Sodium concentrations within other cells—in the skin, muscles, and internal organs—won't have changed, though. The body doesn't like the imbalance. It thinks that sodium concentrations inside and outside cells should be the same. So, by osmosis, the body directs water out of the blood and into the other cells, lowering those cells' sodium concentrations and causing them to engorge. Hands and feet swell. The chest may feel constricted as water fills lung cells. The person gains weight rapidly, since he or she isn't excreting fluids. In extreme cases, the brain swells, leading to disorientation, fainting, coma, even death. That is reportedly what killed the Chicago runner. Hyponatremia has caused deaths at a number of marathons since. Interestingly, during the same time frame, no deaths have been linked to dehydration at marathons.

"Every one of those deaths was preventable," Dr. Noakes says. "If people would have listened to their own bodies and not to the quote-unquote 'experts' advice, they would have survived." They should, he says, have listened to their thirst.

Let Thirst Be Your Guide

The body's ability to regulate its internal water balance is, the latest science tells us, wonderfully precise, if we leave it alone to get on with things. A 2010 review of scientific knowledge about hydration pointed out that most healthy people replenish the fluids that they lose over the course of a day with almost uncanny accuracy, without the need for guidelines telling them when and what to drink. Similarly, the Kenyan runners, although they drank very little during their workouts and lost several pounds each time, rehydrated every evening with such precision that they turned up the next morning weighing the same as the day before and with no clinical signs of dehydration.

The human body's hydration system is, when all is working well, dictated by thirst, which is itself a response to changes in sodium levels. If the sodium concentration in your bloodstream rises even slightly, meaning that your blood becomes less dilute and sludgier, alarm bells ring system-wide and the thirst mechanism is activated. "You don't have to stay ahead of your thirst," Dr. Noakes says. "It's the best indicator there is about the body's hydration status."

But many of us are so studious about drinking fluids all day long that we never feel thirst. A recent survey of recreational runners conducted by researchers at Loyola University found that more than a third of them said that during both training and races they typically drank according to a preset "schedule" of some kind, regardless of whether they were thirsty, and almost 10 percent reported that they believed that for good health and athletic performance they were supposed to "drink as much as possible."

"Plenty of people still believe that eight-glasses-a-day nonsense," Dr. Noakes says, although a comprehensive 2005 report from the Institute of Medicine, based on years of study and scientific review, concluded that there was no credible evidence for the widespread belief that we need to drink eight glasses of water a day, especially since a large portion of a person's daily fluid needs can be met through food. Meat and vegetables contain water. So does milk. Even coffee counts toward your hydration intake, since, although it is a diuretic (meaning it promotes urination), you retain more of a latte's fluids than you lose. Drinking an additional eight full cups of water and other fluids "is likely to be more than most people need," Dr. Coyle says, even if they're quite active.

Certainly the amount of fluids that some people drink during exercise far outstrips their sweat losses. According to a cautionary report in *Medicine & Science in Sports & Exercise*, some back-of-the-pack marathon racers swallow as much as six or seven liters (more than two hundred ounces) of fluid during the twenty-six-mile event. The top runners, meanwhile, drink barely two liters, or just over sixty-seven ounces. "The most dehydrated runners in any race," Dr. Noakes says, "are the winners."

Weigh the Consequences

So how much and what should you drink if you regularly exercise? "That's what we'd all like to know," Dr. Coyle says. On a practical level, the answer seems to depend, as it so often does, on who you are and how you exercise. Going for a mild forty-five-minute bike ride on a cool day? You're not going to sweat much, even if you're the ripest of males, and you won't need to worry unduly about replenishing fluids.

The most widely accepted DIY method of deciding if you're drinking properly is "to weigh yourself before and after a workout or race," Dr. Coyle says. A 2008 study found that most of the other more formal methods of assessing dehydration after strenuous exercise—having dry skin, sunken eyes, the inability to spit, dark-colored urine, and dry mucous membranes—were poorly correlated with whether someone was clinically dehydrated. Only significant weight loss was a good indicator. So weigh yourself before your next hour-long run, bike ride, or other exercise session, preferably without your shoes or shirt on. Drink as you normally do. Weigh yourself again at its conclusion. "If you're not losing more than two pounds of body weight in the course of an hour due to sweating," Dr. Coyle says, "you're fine. Don't change anything. If you are losing more than two pounds in the course of an hour, you may need to drink more." And if, as happens rarely, you actually gain weight, cut back dramatically on how much you drink in subsequent sessions. Weight gain during exercise is the primary symptom of hyponatremia (which, perhaps thankfully, is very uncommon in exercise lasting less than three or four hours and conducted at a slow pace).

Hot and Wet

It's probably important to mention at this point that, contrary to popular belief, being well hydrated will not protect you from overheating. "It's a myth that you have to be dehydrated to develop heatstroke," says Douglas Casa, Ph.D., a professor of kinesiology at the University of Connecticut who's extensively studied athletic performance in the heat, inspired in part by a heatstroke he suffered as a teenage runner. At a triathlon several years ago in Melbourne, Australia, which was mounted on an oppressively hot, humid morning in December, the antipodean version of June, fifteen competitors had to be treated for heat illness. Three of them were diagnosed with life-threatening heatstroke. All were stricken early in the race, before they'd had time to become dehydrated.

More definitively, when South African researchers let competitive male cyclists drink as little or as much fluid as they wanted during a series of grueling time trials in hot, humid lab conditions, most drank less than they sweated away. They lost weight. But they didn't become seriously overheated. In general, the scientists concluded, drinking when they felt like it seemed to be the "optimal" hydration strategy, "as it prevents athletes from ingesting too little or too much fluid."

"The lesson is not 'don't drink during exercise,'" says Dr. Noakes, one of the study authors. "The lesson is 'drink what you need but no more and no less.'"

As to what to drink, water has a long and impressive track record. But sports drinks have their role. "Some people enjoy the taste, so are more likely to drink than with water alone," Dr. Casa says.

The sweeteners in sports drinks also offer a source of carbohydrates, although probably not enough for serious athletes and possibly too much for everyone else. "There are lots of people out there who regularly walk for thirty minutes and then drink a bottle" of a sports drink, Dr. Noakes says, "which contains more calories than they just burned, and then they wonder why they don't lose weight." As for the electrolytes added to the drinks—usually sodium and potassium—they are unnecessary for most healthy people. "We all live in an excess of salt," Dr. Noakes says. Additional salt is rarely needed, no matter how much you sweat or how salty your perspiration may be. Potassium losses in sweat, meanwhile, are almost nonexistent. But if you suspect that, for some reason, you would benefit from more electrolytes, there's always weak, milky bush tea. In the Kenyan study, that concoction was found to contain just enough trace amounts of sodium and potassium to replace the little of those substances that the runners lost.

After the Workout Is Over

Done now with your run, bike ride, grinding tennis match, or Stairmaster session? Many people might expect that sports nutritional concerns would also then end. Those people would be wrong.

Now you need to replenish. "Back in the early nineties, most athletes, especially runners and cyclists, were preoccupied with carbohydrates during exercise," says John Ivy, Ph.D., a professor of kinesiology at the University of Texas at Austin and one of the pioneers of research into the timing of sports nutrition. But eating or drinking only carbohydrates during a long workout rarely fully replaced what was burned, Dr. Ivy says. Over time, athletes could drain their muscles of stored fuel and find workouts harder and harder to complete.

But Dr. Ivy, who also studies blood sugar mechanics in diabetics, noticed that diabetics had better blood sugar control after exercise. "Exercise makes your muscles more responsive to insulin," he says, "and this insulin, in turn, increases glycogen muscle uptake." In other words, exercise prompts your muscles to absorb more sugar from the bloodstream. Your body is primed by the exercise to help itself replenish lost fuel. So it's ideal to eat or drink some carbohydrates now, to increase the levels of sugar in your blood.

This improved insulin response lasts only for a brief time after a workout, though. "You have a window of about thirty to forty-five minutes," Dr. Ivy says. After that, muscles lose some of their acute insulin receptivity and grow less able to absorb sugar.

You can prolong your window of insulin sensitivity if you swallow protein as well as carbohydrates immediately after exercise. "Protein co-ingestion with carbohydrates can accelerate muscle glycogen repletion by stimulating endogenous insulin release," says Luc van Loon, Ph.D., an associate professor of human movement sciences in the Netherlands and the author of many studies of sports nutrition. Translation: Coupling protein with carbohydrates prompts your muscles to store even more glycogen for use during your next workout. Protein eaten soon after a workout may also aid in muscle recovery, if the exercise has been especially prolonged or intense.

Note that this effect is seen primarily with protein consumed after exercise, not while you are still working out. Although some studies have suggested that ingesting protein during exercise might somehow improve performance or aid in recovery, most physiologists are skeptical and the majority of experimental results not supportive. In one representative study, experienced cyclists drank a normal, carbohydrate-only sports drink or one enriched with whey protein during a long easy ride followed by a strenuous hour-long time trial. The researchers found no benefit from the protein. The cyclists didn't produce more power after drinking it, cover more miles during the time trial, or have fewer markers of muscle damage the next day. The protein didn't hurt their performance, but it didn't help in any measurable fashion, either. As Dr. Jeukendrup, who coauthored the study, points out, "There's really no good reason why protein would work" to improve performance during exercise. "Protein is a poor fuel."

But it can do fine things for your muscles afterward. And by a coincidence happy for those of us who already wallow in the stuff, there is one particular food that happens to contain an almost ideal ratio of sugars and protein to spur glycogen replenishment.

That food is low-fat chocolate milk. Science urges you to drink it after exercise.

Those two sentences make me happy.

"I'm a big fan of chocolate milk," Dr. Ivy says. "I drink it all the time."

You could, if you distrust the simplicity of Hershey's syrup, invest in premixed, bottled protein-carbohydrate drinks or prepared sports protein shakes and powders instead. But, really, why?

Voodoo Nutrition

Ultimately, sports nutrition, whether it concerns food or fluids, does seem to boil down to the obvious. "People think there's some voodoo to sports nutrition, some magic formula. But common sense is the most important ingredient," says Allen Lim, Ph.D., a physiologist and trainer who's worked with some of the best bike racers in the world. "If you're hungry, you need to eat. Food happens to be good for that." A typical menu, in his experience, for some of the world's best and most preternaturally active cyclists, he says, might be "lasagna for dinner and muesli or a breakfast burrito for breakfast. Lots of rice. Chicken. Eggs. Sports nutrition is just nutrition, with an extra word in front."

That practical wisdom will not, of course, prevent the continued rise of gimcrack dietary fads. They will stubbornly come. And go. Fat-loading had a brief and greasy heyday, as did the glycemic index diet. Neither withstood scientific scrutiny, particularly fat-loading. The theory was that your body stores far more fat than carbohydrates (even the most skeletal of marathoners has some lard in him) and if you could train your system to preferentially burn fat before carbohydrates, you'd be able to exercise longer and more intensively. Believers cut back dramatically on their carbohydrate intake and substituted fried chicken and buttered everything, and in preliminary studies did burn more fat as fuel during exercise than those on high-carbohydrate diets. But they almost never actually performed better. In a representative study from New Zealand, twelve competitive bike riders ate a high-fat or high-carbohydrate meal before completing a grueling ride. The highly fattened riders burned more fat during the session. But they didn't ride faster or with more power. Similarly, in a slightly nauseating study out of Australia, competitive cyclists were given a high-carbohydrate, normal, or extremely fatty meal a few hours before a strenuous ride. The riders assigned to the high-fat meal received an additional injection of a substance that increases the release of fatty acids into the bloodstream. Their muscles were drenched with fat. During a subsequent strenuous ride, the cyclists given carbs beforehand burned carbs; those oozing fat burned fat. But none rode farther or faster than the others.

Changing an athlete's diet doesn't change the athlete much.

That point is especially true of supplements. Athletes and everyday exercisers tend to be among the world's most eager consumers of pills, shakes, and nutritional nostrums of all kinds. Professional football players have been known to swallow hundreds of different vitamins and other supplements every day, many of them with unpronounceable names and questionable utility. And at a recent American College of Sports Medicine meeting, scientists from Taiwan reported that chicken essence with ginseng has become a popular supplement in that nation; it is thought to speed recovery after prolonged exercise, and in a small-scale study the stuff did exactly that.

More commonly, however, the effects of supplements on health or athletic performance are equivocal, counterproductive, or simply imaginary. In an interesting new experiment, a group of college students underwent exercise testing and then received pills that they believed to be multivitamins. Some pills did contain vitamins, but others were placebos. The students swallowed their assigned pills for three weeks and then repeated the exercise test. All performed better, whether they had received vitamins or not. A control group that took no pills did not improve.

More dramatically, researchers in Germany recently enrolled a group of forty young men in a monthlong exercise program. The men, most in their mid-twenties, were healthy but not athletes, although some were moderately active (playing sports or working out a few times a week) before the experiment began. The rest were sedentary.

The supervised exercise program they undertook for the study was straightforward: twenty minutes of stationary bicycling or treadmill running, followed by forty-five minutes of circuit weight training, during which they moved from machine to machine rapidly enough to keep their heart rates elevated. The men worked out five days a week for four weeks.

Meanwhile, half of the group concurrently began swallowing

antioxidant supplements composed of moderate doses of vitamins C and E. Most of us have heard that exercise increases levels of free radicals and oxidative stress. Free radicals (also called, unpoetically, reactive oxygen species) are molecules created by the breakdown of oxygen during metabolism. All of us constantly create free radicals by the mere act of breathing, so some amount of the circulating molecules are normal, an indication that you're alive. But too many washing around in your system can be dangerous. Since they're highly reactive, they tend to attack other cells and damage tissues, possibly contributing to aging and the development of various diseases. Usually, the body's innate defenses defuse the molecules and keep any damage to a minimum. But if free radical numbers burgeon wildly, the body's defenses become overwhelmed and you develop oxidative stress, a perilous buildup of free radicals.

Exercise causes a surge in the production of free radicals, because we breathe more during exercise. So, some years back, many experts began urging the fitness-minded to pop large doses of antioxidant vitamin supplements, on the theory that if working out led to the creation of extra free radicals, you should take antioxidant pills to sop up those extra molecules, ward off the undesirable oxidative side effects of exercise, and leave yourself that much healthier. Certainly, that was the outcome that researchers expected to see in the German study.

And at first that's what happened. The men taking the antioxidants did display higher blood levels of antioxidants in their bloodstreams than the men who didn't swallow the pills. Those circulating antioxidants might have been expected to fight off the exercise-induced free radicals. But the men's muscles also showed significantly less of certain genetic and enzymatic activity that would indicate that their innate antioxidant defense system was ramping up operations. In effect, the men's bodies noted the high levels of antioxidants in the bloodstream and decided that the innate antioxidant response could go off-line; it wasn't needed.

On the other hand, the men who were not taking the antioxidants had developed stronger, more active antioxidant defense systems in response to the influx of free radicals from their exercise.

What was more puzzling was that the men taking the vitamin pills did not improve their insulin sensitivity, while the men not using the supplements did. In general, moderate exercise would be expected to significantly improve muscles' ability to respond to insulin and draw sugar from the blood for fuel. That's one of the hallmark health improvements associated with being active and the rationale for using exercise to combat type 2 diabetes. But in this case, removing the necessity for the body to deal on its own with the free radicals apparently also prevented other adaptations that make exercise healthy. Antioxidant supplements "prevent the induction of molecular regulators of insulin sensitivity and endogenous antioxidant defense by physical exercise," the German scientists concluded. Or, more bluntly, the vitamins undercut the exercise's benefits.

The German researchers weren't alone in coming to this conclusion. A few years earlier, researchers from the University of Valencia, in Spain, and the University of Wisconsin, Madison, had had laboratory rats run on small treadmills until they were exhausted. Some of the rats had been injected beforehand with a powerful, pharmaceutical-grade antioxidant compound that effectively halts the production of almost all free radicals, even during exercise. After the rats ran, the researchers measured the levels of a number of substances in their leg muscles. Not surprisingly, the injected rats showed almost no free-radical activity.

The leg muscles of the other exercised rats, though, teemed with free radicals. At the same time, they fizzed with other, apparently related biochemical reactions. Genes were being expressed that activated growth factors that, in turn, increased levels of "important enzymes associated with cell defense" and "adaptation to exercise," the researchers wrote. The rats with low free-radical levels exhibited almost no comparable activity. Somehow, the researchers speculated, the free radicals were acting as messenger molecules. They were jump-starting the processes that over time would allow the rats' muscles to adapt to exercise. Without enough free radicals in the body's system, there was no "activation" of these "important signaling pathways" and less healthy adaptations to exercise. Looked at in the light of these findings, they wrote, "the practice of taking antioxidants" to ward off the presumed free-radical damage caused by exercise "may have to be re-evaluated."

Of course, some nutritional supplements do make sense, and one of them, as seems only right, is beer, although of the nonalcoholic variety (sorry), according to new research out of Germany. Researchers at the Technical University of Munich approached runners who were training for the Munich Marathon and asked if they would—in the name of science—drink lots of beer. Hundreds agreed, even when told that the beverage would be nonalcoholic, and in some cases a flavored placebo. The runners downed at least a liter of their assigned beverage every day. After the marathon, those drinking the nonalcoholic beer reported far fewer colds and other illnesses than the runners swallowing the placebo and had other indications of better immune system health. That matters, says Johannes Scherr, M.D., lead author of the study, because if a marathon runner's body is less abused after a race and he doesn't develop the sniffles, he presumably can return to training more quickly than he otherwise might have been able to. "It can be speculated that the training frequency could be higher, with shorter breaks after vigorous training sessions" in those drinking beer, he says.

The beneficial effects seem to result from the beverage's rich bouquet of polyphenols, a substance that is known to fight off viruses, Dr. Scherr says. Of course, alcoholic beer is drenched in polyphenols, too—"even more than nonalcoholic beer," he says—but has the signal disadvantage of being alcoholic. "We do not know whether the side effects of alcoholic beer would cancel out the positive effects caused by the polyphenols," he says. "Furthermore it is not possible to drink one to one and a half liters of alcoholic beer per day, especially not during strenuous training." We all knew that, right?

Ultimately, of course, the most comprehensive lesson from the science about supplements and, indeed, sports nutrition in general is that the body remains adept at seeing to itself. It might need propping up during periods of extreme exertion, such as marathon training, in which case you should lay in a case of nonalcoholic St. Pauli Girl. (A friend with wide and hard-won experience of many, many beers recommends it.)

But otherwise, let your body do its work. It knows how. "The body adapts" to exercise "really well, all by itself," says Li Li Ji, Ph.D., a professor of exercise physiology and nutritional science at the University of Wisconsin and coauthor of the rat antioxidant study. "The human body," he says, "is quite a smart machine."

Some Guidelines for Smart, Healthy Sports Nutrition

1. Ask Yourself: Am I an Athlete?

Be honest. If you're not working out for more than an hour a day or at an achingly strenuous intensity, then, really, you're not. You're looking for health and fitness, which is a fine thing, and you can achieve it with a standard healthy diet. If, though, you regularly exercise for more than an hour, you would be wise to eat a relatively high-carbohydrate diet, meaning about 60 percent or more of your daily calories should come from carbohydrates. This is a larger percentage than in the standard recommended daily allowance, which pegs the percentage closer to 55 percent. If you aren't sure of the composition of some of the elements of your diet (is broccoli a carbohydrate? is cheese?), the helpful Nutrient Data Laboratory at the U.S. Department of Agriculture's website can provide guidance. Plug in the name of the foodstuff and get its relevant nutrient and caloric profile.

2. Carry Carbs.

If you are exercising strenuously, bring carbohydrates to consume along the way. Most sports nutrition experts suggest consuming at least 200 to 300 calories' worth of carbohydrates per hour. A typical energy bar contains between 200 and 300 calories; a gel packet, about 100 calories.

3. Real Food Is Fine.

"Athletes often forget how much of a performance boost you can get from real, simple, natural food," says Dr. Lim. "The best sports diet involves eating food with ingredients you can pronounce, the commonsense stuff your mother told you to eat." As proof, pack a few of his homemade sushi rice bars during your next group ride or long, pre-marathon run, in lieu of prepackaged energy bars. Just combine cooked sushi rice, scrambled eggs, soy sauce, and some bacon or prosciutto, press the mixture into a pan, and cut it into bars. Or do as I do and unabashedly aim for sweetness by substituting Nutella and peanut butter for the savory ingredients. Be prepared to share.

4. Avoid Binge Drinking.

Many of us have been told that we should "stay ahead" of our thirst during exercise and drink as much as possible to avoid dehydration. But the latest science and expert guidelines agree that thirst is actually a reliable physiological marker of your fluid needs. If you're thirsty, drink. If you're not, you probably are sufficiently hydrated.

5. Time Meals and Snacks Appropriately.

Although many of us have been told that if we eat in the hour before exercise, our blood sugar will rise too soon and then plummet disastrously during the workout or race, the latest science suggests otherwise. Blood sugar may drop in some people, but it usually levels out within the first fifteen minutes of exercise. So have a banana before you run. Then eat in the hour immediately following a strenuous session. Replenishing lost fuel is physiologically easiest then, especially if you combine carbohydrates and protein.

6. Two Words: Chocolate Milk.

Recent science has shown that low-fat chocolate milk provides the ideal ratio of carbohydrates and protein to boost fuel replenishment after exercise. In one especially noteworthy experiment, athletes who drank chocolate milk after a workout not only recovered more fully than those who drank a sugary sports drink or water, but they also gained more muscle, lost more fat, and developed greater endurance capacity. Plus, compliance with the chocolate milk regimen was excellent.

4

The Losing Battle

What follows is a story of sorrow. It begins at a highly advanced calorimetry laboratory at the University of Colorado School of Medicine, in Denver, where researchers were examining whether fit and/or unfit people were able to burn extra calories after they had finished working out. It seems as if exercise should incinerate calories even after a workout has ended. It's hard work, exercising. But scientists hadn't scrupulously studied the issue in controlled conditions to see if the process occurred or not. So the researchers recruited several groups of people. Some were lean endurance athletes; others were also lean but mostly sedentary; and the rest were, to be frank, fat, and fairly inactive. Each of the volunteers agreed to spend, over the course of the experiment, several twenty-four-hour periods in a special laboratory room—a walk-in calorimeter—that measures the gases a person inhales and exhales and thereby the number of calories he or she is burning. Using various calculations, researchers can also tell whether the calories expended are in the form of fat or carbohydrates, our bodies' two main fuel sources.

Burning more fat than carbohydrates is obviously desirable for weight loss, since the fat being burned comes primarily from body fat stores, and all of us, even the skinniest, have plenty of those. But the broader issue of just how and when we burn body fat is surprisingly tangled, especially when looked at in combination with exercise. Over the years, science has established fairly definitively that regular endurance training increases the ability of the body to use fat as a fuel during exercise. Muscles turn first to carbohydrates as their preferred fuel source, but repeated long runs, swims, or bike rides will teach muscles to use fat effectively also. A marathon runner needs to use some fat, since no one's body contains enough stored carbohydrates to fuel a three-hour or more exercise bout. Meanwhile, despite the popular notion that slow exercise burns more fat than longer, harder bouts, it doesn't. During a leisurely thirty-minute walk, even if 100 percent of the energy you burn comes from fat (which is impossible; some percentage of fuel always comes from carbohydrates), you'll burn fewer fat calories than someone running for ninety minutes, even as that person burns plenty of carbohydrates, too.

But back at the calorimetry lab, the primary question the researchers hoped to answer concerned how bodies would divide up and use calories not just during but also after exercise. Would the athletes—or any of the other volunteers, fit or not—burn extra fat calories after exercising, a phenomenon that some exercisers (and even more diet and fitness books) call "afterburn"?

"Many people believe that you rev up" your metabolism after an exercise session "so that you burn additional body fat throughout the day," says Edward Melanson, Ph.D., a professor of endocrinology at the University of Colorado and one of the authors of the study. If afterburn were found to exist, it would suggest that even if you replace the calories you use during an exercise session—if you have a candy bar after that hour on the treadmill—you still should lose weight, without any additional work on your part, providing the proverbial free lunch.

Each of Melanson's subjects spent twenty-four quiet hours in the calorimeter, without any planned activity, followed later by another twenty-four hours that included an hour-long bout of stationary bicycling. The cycling was deliberately performed at a relatively easy intensity (about 55 percent of each person's predetermined aerobic capacity), in expectation that the low intensity of the exercise would prompt their bodies to burn a relatively high proportion of their calories from stored fat. All of the subjects ate three normal meals a day.

Then the researchers compared the pattern of calorie burning indicated by the calorimetry readings. Did people burn more fat calories in the hours after they'd exercised? The answer was an unequivocal no. It turned out that none of the groups, including the athletes, experienced "afterburn." They did not use additional body fat on the day they exercised. In fact, most of the subjects burned slightly less fat over the twenty-four-hour study period when they exercised than when they did not, presumably because they were burning some carbohydrate calories during their exercise session and not so much when they just sat around quietly.

"The message of our work is really simple," although perhaps not agreeable to hear, Dr. Melanson says. "You don't get any freebie calories." Exercise, or at least moderate exercise, doesn't rev up your metabolism and turn you into a fat-burning furnace for the rest of the day. The extra calorie burning ends with the exercise session. So if you follow up your workout with a Mars bar, you likely will have replenished the calories you just burned (and possibly then some). "It all comes down to energy balance," says Dr. Melanson—or, as you might have guessed, calories in and calories out. People "are only burning two hundred or three hundred calories" in a typical thirty-minute exercise session, he points out. "You replace that with one cookie."

"Exercise Is Useless"

Why exercise doesn't inevitably make people skinny is one of the more intriguing and vexing issues in physiology. Yet study after study finds that, with some rare exceptions, it does not. In a typical report out of Australia, fifty-eight obese men and women completed twelve weeks of supervised aerobic training without changing their diets. The group lost an average of about seven pounds, quite a bit less than would have been expected, based on how many calories they were expending during the exercise. Many lost barely half that. A few gained weight. All of which is common. According to a recent comprehensive review of decades' worth of studies of weight loss and supervised aerobic exercise programs, the average participant lost barely a third as much weight as researchers had expected, and in almost every instance, some of the volunteers wound up fatter.

"In general, exercise by itself is pretty useless for weight loss," says Eric Ravussin, Ph.D., a professor at the Pennington Biomedical Research Center and an expert on the topic.

That harsh reality is familiar to many of us. "I doubt if there's an exercise scientist who hasn't been cornered at some point by someone who wants to know why, if she's going to the gym faithfully three or four times a week, she's not getting any thinner," says Barry Braun, Ph.D., a professor of kinesiology at the University of Massachusetts, Amherst, and director of its Energy Metabolism Lab, who has extensively studied the issue of exercise and weight loss.

It's a fair question. Why *doesn't* exercise slough away flab? It burns calories, after all, including fat calories. That's not in doubt. "Walking, even at a very easy pace, means that you'll probably burn three or four calories a minute, beyond what you would use quietly sitting in a chair," said Dan Carey, Ph.D., a professor of exercise physiology at the University of St. Thomas, in Minnesota, who studied the mechanisms of fat burning. In general, the mathematics of weight loss, whether achieved by exercise or by any other means, is uncomplicated, involving only subtraction. "Take in fewer calories than you burn, put yourself in negative energy balance, lose weight, the end," says Dr. Braun, who has conducted multiple studies of exercise and weight loss over the years.

The caloric deficit can result from cutting back on how much you eat or from increasing how much you burn (i.e. increasing your activity level), or both. Your body doesn't care or, by and large, notice any difference. That fact was apparent in a study recently completed by Dr. Ravussin and his colleagues, in which a group of overweight adults began a six-month weight-loss program. Some cut back on the total number of calories they ate each day by 25 percent. Others cut their calories by 12.5 percent while increasing their activity by 12.5 percent, so that their overall energy balance was, likewise, negative 25 percent.

At the end of the six months, there was no difference in the weight loss (minor, in either case) between the two groups, which seems encouraging. But there are serious questions about whether the exercise program would be sustainable over a longer period—or when researchers weren't supervising most sessions, as they had been. The "dose" of exercise required was nearly an hour a day, five times a week, of moderate-intensity activity, "which is a lot more than what many people would be able or willing to do, especially without someone tracking them," Dr. Ravussin says. "This idea that you can walk your way to fitness, it's just not proving to be true."

In fact, it may be that, in many instances, the harder someone tries, the more likely he or she is to undercut her own efforts. In a telling recent study, researchers asked more than 400 middle-aged women to begin a series of divergent exercise routines. Some worked out at a gentle pace (walking or riding a stationary bicycle) for a total of 72 minutes a week, which is about half of the recommended amount for health benefits. Another group met the current national exercise guidelines for health, exercising easily for 136 minutes a week. The final group exercised at a volume 50 percent greater than the guidelines suggest, for the equivalent of almost 45 minutes a day five times a week (although some lumped their sessions into two long workouts a week). All were asked to maintain their old eating habits, although their diets weren't supervised. At the end of the six months, both the lightest-exercising group and the group that had met the current exercise guidelines had lost weight, as researchers would have expected based on the energy they were expending. This didn't translate into a lot of weight loss; the light-duty exercisers only shed about three pounds, on average, over the six months. But the most diligent exercise group lost even less, barely dropping two pounds, although they would have been expected to lose closer to eight. So the group that had exercised the most had lost the least. "Our findings provide an excellent example of the complex nature of the energy expenditure and energy intake relation," the authors drily understated.

The Compensators' Diet

So what is going on? Why does exercise have so little impact on weight loss for so many people? The answer is almost certainly, as a number of the new studies say, "multifactorial," and one of the most reverberant and bizarre of those multiple factors is that, as a species, we're astonishingly efficient at clawing back lost calories, or, in scientific parlance, compensating.

"People's systems are pretty good at recognizing when they have entered negative energy balance and doing something about it," Dr. Ravussin says, "which usually involves regaining weight."

Compensation can take many forms, but in the context of exercise and avoirdupois, none is more devious than what has been called "non-volitional exercise-induced inactivity." People relying on exercise to burn calories sometimes, without deliberate intent, begin moving less during the rest of the day. They grow more sedentary. They sit when they once might have stood or walked around the house. A similar dynamic occurs when you reduce food intake. The body tries hard to maintain its weight. In a study of obese rats whose chow calories were slashed by 50 to 60 percent for four months, the animals became markedly less active, even after they were allowed to return to their normal, plentiful diet. They continued to slouch in the corners of their cage at every opportunity, except when they skittered to the food tray. Within two months, they'd regained every ounce.

Human studies have produced similar, if less creepy results. In one study, 34 overweight women completed an eight-week exercise program that consisted of 150 minutes per week of moderate stationary bicycling. They kept detailed activity and diet diaries, beginning the week before they started exercising. By the end, only 11 had lost the expected amount of weight (an average of 2.2 pounds). The other 23 had not. One had gained five pounds. When the researchers carefully parsed the women's diaries, it became clear that many of them were eating more (despite being told not to change their diets) and many more were moving less. They were taking significantly fewer steps each day, not consciously, but consistently. Interestingly, the 11 women who met the weight-loss expectations were, on average, moving more outside of the exercise sessions than they had been before the program began.

Sloth, however, is not the only weapon our bodies deploy to undermine any walk-(or jog or bike)-to-weight-loss efforts. The body also recalibrates appetite.

The Unfairness of Leptin

Not long ago, Dr. Braun oversaw a study in which eighteen overweight men and women walked on treadmills in multiple sessions, after which they were directed to eat either as many calories as they had just burned or fewer (meaning they went into negative energy balance). The experiment sparked notable alterations in the volunteers' blood levels of various hormones that control appetite, which was to be expected. In general, the mechanisms underlying appetite and energy balance in the human body are neatly regulated. "The body aims for homeostasis," Dr. Braun says. It likes to remain at whatever weight it's been used to. So small changes in energy balance can produce rapid changes in the hormones most associated with appetite, particularly acylated ghrelin, which is known to increase the desire for food, as well as insulin and leptin, hormones that affect how the body burns fuel.

After each session, the volunteers' blood was drawn to test their appetite-hormone levels. Overall, the men displayed little or no changes in their energy-regulating hormones or in their appetites, whether they were in negative energy balance or not. But the women uniformly showed increased blood concentrations of acylated ghrelin and decreased concentrations of insulin after the sessions in which they had eaten less than they had burned. Their bodies were directing them to replace those lost calories, immediately.

In physiological terms, the results are "consistent with the paradigm that mechanisms to maintain body fat are more effective in women," Dr. Braun and his colleagues wrote. In practical terms, they're scientific proof that life is unfair. Female bodies, inspired almost certainly "by a biological need to maintain energy stores for reproduction," Dr. Braun says, fight hard to hold on to every molecule of fat. Exercise for many women (and some men) inexorably increases the desire to eat.

Burn, Baby, Burn

There are exceptions, of course, and they tend to arise from exercise that is strenuous and repeated, as you might have guessed from observation. The world is not full of chubby marathon champions. The latest science suggests that prolonged or painfully intense exercise can suppress or, at minimum, not initiate increases in appetite. It may even affect afterburn.

An interesting recent review of dozens of studies of the effects of strenuous workouts on appetite found that, in general, exercising for more than sixty minutes or at an intensity biting enough to incinerate eight hundred calories led to decreased appetite in people, particularly men. These types of workouts, the review found, tended to increase the concentration of leptin in the blood, which "promotes lean body mass by acting on receptors within the hypothalamus to decrease food intake." This finding was borne out in a study presented at a recent annual meeting of the American College of Sports Medicine, in which fit, healthy young men ran vigorously for an hour and a half on a treadmill. Afterward, their blood concentrations of acylated ghrelin had fallen, their levels of leptin had risen, and food held little appeal for the rest of that day. Exercise had blunted their appetites.

The fit and lean find it relatively easy to become fitter and leaner.

Even the otherwise chimerical afterburn may be ignited if you sweat hard enough, according to at least one study. When the country's newest metabolic testing chamber was completed on the grounds of the University of North Carolina at Chapel Hill not long ago, local physiologists promptly began monopolizing it to test the effects of exercise on metabolism. (Physiologists, who usually exercise regularly, are keenly interested in how workouts influence physique.)

One of those scientists was David C. Nieman, Ph.D., a professor at Appalachian State University and the author of innumerable studies of exercise. He was an afterburn skeptic. In past editions of his well-known textbook, *Exercise Testing and Prescription*, he'd written that the available evidence suggested that exercise, no matter what its intensity, caused little if any metabolic amplification.

But then he and his colleagues were given access to the new UNC metabolic chamber and entrenched notions would fall. He and his fellow researchers began by recruiting a group of healthy young men aged twenty-two to thirty-three. None were athletes, although some, during baseline fitness testing, displayed above-average endurance capacities. A few were overweight. "We wanted a range of endurance levels and body types," Dr. Nieman says.

The volunteers showed up twice for separate, twenty-four-hour sessions in the metabolic chamber. Housed in their own rooms, they spent the first sessions mostly resting—watching television, reading, or otherwise occupying themselves quietly. Throughout that time, the chamber hummed and collected expired gases and calibrated their energy output.

In the second session, they worked out, riding a computerized stationary bicycle for forty-five minutes. The pace was strenuous and draining, amounting to about 75 percent of each volunteer's endurance capacity. At that intensity, you'll sweat and pant and "feel pretty seriously uncomfortable," Dr. Nieman says. "But all of them, even the men who were overweight, completed the workout, so anyone can."

Afterward, all of the volunteers were directed to eat enough additional food to replace the calories that they'd burned during the riding—an average of 519 per man—and rest for the remainder of the day. Again, the calories that they used during their twenty-four-hour stays were enumerated.

The scientists then compared the total caloric expenditures during each session.

To Dr. Nieman's surprise, it turned out that the men were burning more calories after they'd exercised. For fourteen hours after their cycling sessions, in fact, their metabolisms remained in overdrive, even a few hours after they were asleep. At the end of the fourteen hours, the men had burned, on average, 190 additional calories, compared to when they hadn't exercised. They hadn't moved about more during their quiet hours after exercising, but their metabolisms, apparently revved by the pedaling, had chewed through more energy.

The implications of this finding are rousing but also extremely limited. "I think there is good news" in the data, "in terms of the potential effects of exercise" on weight loss and control, Dr. Nieman says. Together, the 519 calories burned during the actual cycling session and the 190 additional calories accrued afterward "represent enough calories" to lead to weight loss over time—if you do not immediately replace them, plus more, which, as we now know, is distressingly easy to do. Those 190 effortless afterburn calories are fewer than are found in a thirty-two-ounce bottle of Gatorade or a single Milky Way. Also, Dr. Nieman and his colleagues had their volunteers replenish the calories lost during the exercise bout. Would not replacing them have affected results? It's impossible to say but plausible. The body, fearing that this sudden disappearance of sustenance signaled famine, might slow its metabolism, ramp up production of acylated ghrelin, or otherwise sabotage that hard effort.

Other obvious and important questions also remain unanswered, particularly related to gender. "This study does not tell us anything about the effects of exercise on other people" who are not young, healthy, and male, Dr. Braun says. "Whether there will also be an elevated postexercise metabolic rate in, for instance, women" who dutifully pedal themselves into a lather "is unknown."

Still, the findings are suggestive. And what they suggest is that to get weight-loss benefits from exercise, you need to push yourself. "We've become a nation of exercise wimps," Dr. Nieman says. "Too many people don't bother or are afraid of exercising hard. But intensity is probably the only way to lose weight with exercise."

Maintaining Hope

Thankfully science does provide more encouraging news about how exercise can combat fat gain, including among women and those who aren't annoyingly fit to start with. For one thing, regular activity, new data suggest, may subvert a potentially fat future. Several years ago, researchers discovered that people who carry certain variations of a gene known as the fat-mass and obesity-associated, or FTO, gene have an enormously increased risk of becoming obese over their lifetimes. Close to a third of Americans of European descent may harbor this gene. But according to a major, ongoing study in Europe, activity begun during adolescence seems to "overcome the effect" of being an FTO carrier. In the study, teenagers whose genotypes were found to contain a version of the FTO gene were weighed, measured, and asked about their daily activities. Those who professed to be physically active for at least an hour a day on most days of the week had significantly lower body mass index numbers than the young FTO carriers who did not exercise, many of whom were well on their way to obesity.

Exercise also lessens the chance that high-fat foods will translate into fat-marbled thighs or, more important if less aesthetically apparent, fat-clogged arteries. An ongoing, large study at Johns Hopkins University persuasively suggests that for anyone contemplating an Atkins-style, low-carb/high-fat diet exercise is essential.

Such diets can be effective, research shows, at helping people to shed pounds. "In our research, it took subjects less time to lose ten pounds" on a high-fat diet than on a high-carbohydrate, low-fat diet constructed using guidelines from the American Heart Association, says Kerry J. Stewart, Ph.D., director of clinical and research exercise physiology at Johns Hopkins University School of Medicine and lead author of the report.

Those following the Heart Association diet required an average of seventy days to lose ten pounds. The high-fat dieters met that goal in about forty-five days.

But a unique worry raised by the high-fat, low-carbohydrate regimen, extending back to the heyday of the original Atkins diet in the 1970s, is that indulging in repeated fatty, glistening meals will lard your arteries and cause heart disease. Some past studies of Atkins dieters found that they were more likely to develop cardiovascular disease than people following other diets.

In one particularly telling study, sedentary people on an Atkins-style low-carbohydrate diet lost quite a bit of weight in the first six weeks of dieting. But almost none of those pounds came from their midsections. Their waistlines didn't budge, and, apparently as a consequence, they showed early signs of impaired blood vessel health after a month and a half on the diet. Their heart disease risk had risen. There seems to be something about fat around the waistline that negatively affects heart health, even if someone loses weight elsewhere on his or her body by following a high-fat diet, according to Shane Phillips, a professor at the University of Illinois at Chicago who conducted the study.

But his subjects had not exercised.

Hoping to determine whether some sweat with their bacon might affect high-fat dieters' heart-disease risk, the Johns Hopkins researchers recruited a group of forty-six healthy but overweight men and women and randomly assigned half to a high-carbohydrate, low-fat, American Heart Association–approved diet consisting of fruit, grains, vegetables, and low-fat meat. The other twenty-three volunteers were assigned to a meatier, cheesier, high-fat, low-carbohydrate spread, with about 55 percent of calories derived from fat; the diet avoided trans fats. Both approaches reduced volunteers' normal daily caloric intake by about 750 calories.

At the same time, the volunteers began a moderate exercise program consisting of brisk walking or jogging and weight lifting. The sessions lasted for at least thirty minutes three times a week and were supervised.

At the start of the experiment, all of the volunteers had healthy blood vessels, according to the widely used blood-pressure cuff test. In it, researchers tighten a cuff around a volunteer's arm, then release it and track the resulting gush of blood to the volunteer's fingertips. During such a surge, healthy vessels dilate, or relax, but unhealthy ones stiffen and narrow, impairing blood flow and indicating possible incipient heart disease.

The researchers then sat back and waited for each person to lose ten pounds, after which they were retested with the blood-pressure cuff. Not surprisingly, the group eating the American Heart Association–style diet continued to display normal, healthy blood flow. But so did the high-fat, intensively meaty diet group. Their blood vessels dilated just as well as those eating a lower-fat diet. "There was no evidence of any harmful vascular effects from the low-carb diet," Dr. Stewart says.

Better, the volunteers had become physically more fit during the experiment, increasing their endurance capacity, a change that, in and of itself, is associated with lower cardiovascular disease risk. They also, unlike volunteers in the earlier studies of sedentary Atkins dieters, had lost weight around their middles.

Of course, the experiment tracked them and the state of the blood vessels for only a few weeks. "It's true that those on the high-fat diets showed no harmful impacts after forty-five days or so" if they exercised, says Dena Bravata, M.D., an internist at Stanford University who has conducted studies of the health impacts of different types of diets. "But what about in five or ten years, if they remain on the diet?" Will jogging three days a week be able to mitigate the effects of years of consuming hefty portions of butter and red meat? At the moment, no one knows.

Still, even the preliminary results strongly suggest that if you are going to adopt a high-fat diet, you should exercise. For his own part, Dr. Stewart has amassed strong and intimate anecdotal evidence in support of that idea. During the small pilot study for his experiment several years ago, he served as a guinea pig, following a low-carbohydrate, high-fat diet and beginning his first consistent exercise routine. In the years since, he has maintained both the diet and the workouts. He is forty pounds lighter these days, he says, and continues to ace tests of his blood-vessel health.

Even those of us whose forays into high-fat meals are far more sporadic—indulging annually in too many Christmas cookies or portions of beef Wellington, for instance—will lessen our chances of negative health consequences if we work out, at least if we can bring ourselves to work out before breakfast. When scientists in Belgium recruited a group of healthy, active young men and stuffed them for six weeks with a breathtakingly lousy diet—50 percent fat and 30 percent more calories than the men had been consuming—most of the brave (and possibly foolhardy) volunteers gained weight. Many also developed insulin resistance and unhealthily fat-marbled muscles, even if they exercised strenuously each afternoon. But not a subset who had been assigned to exercise first thing in the morning, before they had eaten. The exercisers (who ran or bicycled for ninety minutes at a punishing pace) gained almost no weight and showed no signs of insulin resistance. Their bodies, without carbohydrates from breakfast circulating in the blood, turned instead for fuel to the plentiful dietary fats that the men were consuming. "Our current data," the study's authors wrote, "indicate that exercise training in the fasted state is more effective than exercise in the carbohydrate-fed state to stimulate glucose tolerance despite a hypercaloric high-fat diet." If, in other words, you're going to eat badly, exercise can be a palliative, although, as seems only fair, it requires that you drag yourself from bed early and eat no doughnuts first.

In general, in fact, exercise seems to be one of the few reliable means of avoiding stacking on pounds. If it doesn't aid much in weight loss, it is helpful in weight *control*. In a recent, cautionary study, for instance, researchers monitored a group of almost 5,000 young men and women over the course of 15 years. Most of the participants, who ranged in age from 18 to 30 at the start of the study, gained more than a pound a year (a degree of padding that, sadly, is almost standard in America). But those who regularly got out and walked, particularly if they walked for more than 15 minutes a day, tended to gain less, with the most benefits accruing among the people who were heaviest at the start of the study. Obese adults who walked more than 15 minutes per day gained significantly fewer pounds over the 15 years than some of the people who'd been, at the beginning, normal weight but had remained sedentary since.

Researchers from Harvard University found, likewise, when they examined the weight-change histories of more than thirty-four thousand participants in a major, long-term investigation of women's health, that exercise aided greatly in weight maintenance. The women in question had entered the study when they were middle-aged, mostly in their late forties to midfifties, and were followed for thirteen years. During that time, the group gained, on average, six pounds. Some packed on far more. But a small minority gained much less, coming close to maintaining the physique with which they'd started the study. Those were the women who reported exercising almost every day for an hour or so. The exercise wasn't strenuous. It was the equivalent of brisk walking (although some of the women swam or biked or attended dance classes). But the women were diligent. Exercise "wasn't something the women started and stopped," says I-Min Lee, Ph.D., a professor at the Harvard School of Public Health who led the study. "It was something they'd been doing for years, and obviously found benefits in. When you don't gain as much weight as other women, that's a benefit you can see."

Gone for Good

Exercise also, in ways that are only now becoming clear, is essential for preserving weight loss, should you manage to pry off pounds. "When you look at the results in the National Weight Control Registry," Dr. Braun says, "you see over and over that exercise is the one constant" among people who've maintained their weight loss. Close to 90 percent of the people who've enrolled in the registry and who have managed not to regain weight say that they regularly exercise, a result seen in many studies as well. In one representative experiment, nearly one hundred healthy but slightly overweight women began a drastic eight-hundred-calorie diet, continuing until they lost an average of twenty-seven pounds each. Some of the women were then assigned to an easy jogging program or to a weight-training regimen, while others were told not to exercise. All were allowed to eat as much as they wanted.

After a year, all had regained weight. But those who had stuck with either of the exercise training programs for the entire year had regained barely half as many pounds and, even more significantly, had gained almost no weight around their middles, unlike the non-exercisers. It's well known that abdominal (also known as visceral) fat is particularly unhealthy, contributing to metabolic problems, diabetes, and heart disease. "It's quite good news that exercise appears not only to reduce weight regain after exercise, but to keep visceral fat gains to, effectively, zero," says Gary Hunter, Ph.D., a professor of exercise science at the University of Alabama, who led the study.

Scientists still aren't completely sure how and why exercise is so important in allowing people to maintain weight loss. But in experiments with animals, exercise appears to dramatically remodel the metabolic pathways that affect how the body stores and utilizes food. In a fascinating study conducted recently at the University of Colorado, Denver, by Dr. Melanson and his colleagues, scientists fattened a group of male rats. The rodents already had a genetic predisposition to corpulence and, thanks to a high-fat, all-you-can-eat spread, fulfilled that genetic destiny. After sixteen weeks of eating like pigs (or, more accurately, insatiable rats) and lolling around lazily in their cages, they were, by rodent standards, obese. The scientists then switched them to a calorie-controlled, low-fat diet. Reluctantly, since rats enjoy their chow and the richer the better, the animals shed weight, dropping an average of about 14 percent of their body weight.

Afterward, the rats were put on a tightly controlled diet designed to maintain their weight loss. At the same time, half of them were made to run on a treadmill for about thirty minutes a day. The other half remained resolutely sedentary. The runners got to eat slightly more than the other rats, to ensure that they wouldn't continue to lose weight. For eight weeks, they were kept at their lower body weights, in order to establish a new baseline weight.

Then the fun began. For the final two months of the experiment, the rats were allowed to eat as much as they wanted, meaning they could relapse into every old bad eating habit. The rats that had not been running fell greedily upon the kibble. Most regained the weight they originally had lost and then some.

But the exercising rats did not. In a complicated loop, their ratty brains and muscles sent signals to the stomach indicating that the animals were satiated much sooner than the non-exercising rats. They ate less, although they were being allowed to eat as much as they wanted. Perhaps even more important, when the scientists studied their bodies' fuel utilization, it turned out that the exercised rats were metabolizing calories differently. They tended to burn fat immediately after they ate, while the sedentary rats preferentially burned the carbohydrates from the chow. The fat calories in their food were then saved up and stored as body fat. Although the exercised animals did regain some weight, their relapses were far less extreme. Exercise "re-established the homeostatic steady state between intake and expenditure to defend a lower body weight," the study authors concluded. Running had remade the animals' metabolic systems so that they chose to eat less.

Fit, Fat, and Everything in Between

Perhaps the most profound and lingering impact that exercise has in weight-loss or weight-control programs, though, is that it makes you healthier. In the Australian study, during which exercising volunteers achieved much lower than expected weight loss, they also gained other, sometimes ineffable benefits. Each significantly increased his or her aerobic capacity, decreased his or her blood pressure and resting heart rate, and, the authors wrote, achieved "an acute exercise-induced increase in positive mood." They were happier, even if not much thinner. "Significant and meaningful health benefits can be achieved even in the presence of lower than expected exercise-induced weight loss," the authors wrote. Remember that when you stand, disconsolate, on the scale.

Other scientists have found that physical fitness, whether or not it's accompanied by personal fatness, leads to a longer, more robust life. In a landmark study published in 2006, researchers with the Cooper Clinic, in Dallas, reported that among a group of more than 11,300 women who were tested at the clinic and then followed for decades, those in the lowest 20 percent of aerobic fitness were far more likely to die prematurely than the fittest women, no matter what their weight. The researchers had found the same striking relationship between fitness, fatness, and mortality among tens of thousands of men in earlier testing. "Regardless of what fatness category you're in," says Stephen Farrell, Ph.D., the lead investigator for the study, "your risk of premature death is lessened if you're fit."

Which is not to say that you shouldn't continue to try to lose a few pounds if you're heavy, even if you already run, swim, or otherwise work out. Fitness does mitigate but won't alleviate the unsavory consequences of being stout. When scientists pulled data on almost forty thousand women enrolled in the massive Women's Health Study in 2009, they found that those with a higher body mass index, even if they were active, had an elevated risk of coronary heart disease compared with active women of normal weight. Fat cells are not, the study's authors pointed out, benign. They can release inflammatory molecules, which increase the risk of diabetes and heart disease. They can also interfere with muscle function.

So many of us probably need to make some decisions about where we sit on the fat-fit spectrum and if it's a comfortable spot. Science, unfortunately, can't offer much certainty. "This whole fit and fat debate has unbelievable levels of complexity," says Dr. Timothy Church, who's also the director of the Laboratory of Preventive Medicine at the Pennington Biomedical Research Center and has studied exercise and weight loss for years. "One thing we do know, though, is that, no matter what your weight, you should have your metabolic profile tested" to look for signs of incipient insulin resistance or other metabolic perturbations. "Also a waist circumference greater than forty inches for men and thirty-five inches for women is not great. You might want to lose a few pounds in that case," Dr. Church says. "But beyond that, who knows what's normal or ideal? Not me and I study this issue for a living."

Dr. Ravussin, who has frequently collaborated with Dr. Church, agrees, to a point. "Obviously, it's going to be better to be fat and fit than fat and not fit," he says. "But it's even better to be fit and not-so-fat. We all know that, whether we like to face it or not. If you need to lose weight, you have to go into negative energy balance. Eat less or exercise more or both. It's all the same for energy balance, but with exercise you get improvements in your metabolic profile, which is a plus. Then don't go out and eat a lot of food afterward to compensate. It's that simple." And for many of us, it's that exquisitely hard.

Ways to Use Exercise to Control Weight

1. Try Props.

A plethora of studies have found that keeping scrupulous diaries of what you eat and how much you move around can decrease the former and goose the latter, helping you avoid reflexively compensating for lost calories. Pedometers are useful in this regard, since they can track how many steps you take in a day. Many people take fewer after they lose weight, even if they are dutifully exercising. Most of us also overestimate the number of steps we're taking. In one study, people guessed that they were walking about four miles a day during general activities of living (and not as part of their exercise routines). But under the pitiless scrutiny of an accelerometer (an advanced pedometer), it turned out that most were walking less than half that much.

2. Understand the Fat-Burning Zone.

There is a fat-burning workout zone, as many of us have heard. During moderately strenuous exercise, your body turns first to stored carbohydrates for fuel. But at lower intensities—during, for instance, a walk instead of a jog—your body will burn a larger percentage of its calories as fat. This does not mean that you will lose weight, according to Dr. Carey. Recently, he published formulas in the *Journal of Strength and Conditioning Research* detailing the heart rates at which a person maximizes fat burning. "Heart rates of between 105 and 134 beats per minute," Dr. Carey said, represent the fat-burning zone. The problem with exercising in this "fat zone" is that it is not much of a calorie-burning zone. The workload is too slight. In order to incinerate as many calories as you would if you exercised more strenuously, you would have to exercise for a longer period of time. Most of us don't. Also, the fat calories being burned usually come from your most recent meal, not from your love handles. So don't kid yourself about what you're accomplishing with a brief, leisurely stroll. "It's probably best to work out near the top of that heart-rate zone," Dr. Carey said, "so that you burn more calories overall."

3. Push Yourself Sometimes.

Intense exercise is also, it would seem, the only way to ignite a postexercise afterburn of calories, during which your metabolism remains revved up for several hours and you effortlessly incinerate at least some additional calories. Multiple studies have shown that light-duty exercise produces no afterburn. But there is some evidence that long, achingly strenuous workouts do pay off with short-term increases in your metabolic rate. This is not exercise for the fainthearted (or those without clearance from their cardiologist). "The indications are that the exercise must be vigorous and prolonged," says Dr. Nieman, who led the aforementioned study of afterburn. "Walking is wonderful exercise," he says. "I don't want to discourage people from walking. The benefits for general health are huge. But if you expect it" to lead to amplified calorie burning for hours afterward, "you will probably be disappointed." His advice? "For the most bang from your workout" in terms of weight loss, "intensity is the way to go." Run instead of walk, or "put on a weighted backpack and go up and down stairs." And, painful as it may be, skip that celebratory post-workout gelato, unless you are one of those exercisers who worry about keeping weight on, in which case keep your problems to yourself.

4. Work Out Before Breakfast.

It may be unpleasant to roll out of bed and directly into your running shoes, but credible research shows that exercising in a fasted state, even if you exercise intensely, leads to a greater degree of fat burning than when you've eaten first. Again, most of this fat is likely to come from recent meals, not from your fat stores, but the method has been shown to keep people from gaining weight, even on a high-fat diet, and to improve their metabolic profile, almost rendering predawn rising palatable. And when you do eventually get around to having breakfast, perhaps include eggs. Other research has shown that healthy people who consume eggs for breakfast down fewer calories in the remainder of day than those enjoying a bagel or other high-carbohydrate foods.

5. Stand Up and Cheer.

"Emerging evidence suggests that, unlike bouts of moderate-vigorous activity, low-intensity ambulation, standing, etc., may contribute to daily energy expenditure without triggering the caloric compensation effect," Dr. Braun wrote in a recent American College of Sports Medicine newsletter. This means you can lose weight by standing up as you read this. In research at Dr. Braun's lab, volunteers sat for an entire day in a wheelchair. In another session, they stood all day, "not doing anything in particular," Dr. Braun says, not hopping or running or "exercising" in any way. Just standing. The difference in energy expenditure among the same people in the two sessions was remarkable, representing "hundreds of calories," Dr. Braun says, with no increase in the standing volunteers' levels of hormones that stimulate appetite. The standees burned calories but weren't motivated to eat more. So if you want to lose weight, Dr. Braun says, "you don't necessarily have to go for a long run. Just get rid of your chair."

5

What We Mean When We Talk About Endurance

I began running when I was in college. I'd been a track athlete in high school, but that didn't mean that I ran. By the time I was in high school, Title IX was in effect, assuring equal access to organized sports for girls, but its impact was slow to reach the suburban Midwest, where I grew up. There was no girls' cross-country team at our high school, since cross-country courses were two or three miles long, and, at that distance, a girl's uterus could fall out. As for track and field, the boys' team was large and well funded, but our side was tiny, with our sole coach being a bored football assistant who had time on his hands in the spring. After school, my fellow female track athletes and I would change into baggy gym shorts and sneakers, troop out to a sticky asphalt oval under a baking, high-arched sky, and, following Coach's terse instructions, sprint until we vomited.

I quit track after my sophomore year.

But in college I discovered running, not as part of a team but in solitude. From my dormitory, you could jog a few blocks and arrive at the agricultural department's experimental fields. I liked to circle them at a slow trot. They smelled sweet and loamy and sharply chemical. (I was an English major, untroubled about pesticide off-gassing.) Some fields sprouted low vines or flowers; others, tall green leafy plants. Trees edged the fields. It was a quiet place and I usually had the dirt margins to myself.

At first, when I'd reach the fields, huffing, I'd slow and walk around them and then jog home. I don't know the distance, but I'd guess that a lap around the fields was half a mile or so. I wasn't tracking mileage; I just itched to get outside and move after six hours of reading *Coriolanus* or *The Golden Bough*. Plus, being eighteen and a girl, I didn't want to get fat.

But—stop me if you've heard this before, and if you're a runner, you have—it turned out that I liked to run. After a few months, I'd reach the fields, breathing easily now, and keep going, three, four, five times around the plowed rectangles, and then, still un-tired, run home. In the last few blocks, I'd sometimes stretch into a sprint. Without my telling them what to do, my knees would rise and my arms would pump. I'd toe off hard, feeling powerful and fast. I'd never been fat as a kid. My family tended toward the scrawny. But I hadn't been athletic, either. My track career had been, as already noted, ugly, abbreviated, and wet. My steadiest athletic success had come in fourth-grade softball, when I was short, with no discernible strike zone, and drew frequent walks.

But now suddenly I had physical competence and even grace. My thigh muscles bunched and lengthened with unselfconscious animal beauty. Who knew they could do that? When a boy from my statistics class suggested we spend time together, I asked if he'd like to run. He lasted two blocks. We didn't date again. Who cared? There was a brief moment, during one especially gratifying run in my freshman year, as I powered past my dorm and kept going, when I thought about switching majors to biology and premed, the better to bore into the operations of respiration, muscles, and the heart that I could feel at work beneath my skin. But those fields required too much math. Even more, running had poetry, and so it was part of my field. "Good for the body is the work of the body," Henry David Thoreau wrote. "Good for the soul the work of the soul, and good for either the work of the other." And that sentiment, I felt even then, is what we mean when we talk about endurance. That and, of course, aerobic glycolysis.

Staring a Mile into the Future

It's often called "cardio," which makes it sound clinical or medicinal. But aerobic exercise—endurance—is the wellspring of fitness. In more poetic and literal terms, cardio is exercise's beating heart, and its value is inestimable. A growing body of science suggests that aerobic fitness may be the single most important determinant of how long you live, trumping whether you smoke or are obese. A major new study of almost 10,000 American men aged 20 to 82, for instance, found that, over the course of 5 years, those who were the least aerobically fit were the most likely to die of all causes and at all ages. The one-fifth who were the most fit were the least likely to die of any cause, particularly among the 80-plus contingent. Similarly, in a 15-year study of more than 2,000 Norwegian middle-aged adults, those who improved their fitness during the study, no matter how slightly, wound up with a "significantly lower risk of all-cause mortality," even if they were, to be unkind but accurate, fat.

Perhaps most telling, a Cassandra-like recent study undertaken in part by the Cooper Institute showed that the speed with which a man or woman can run a mile in his or her 40s or 50s almost eerily predicts heart disease risk 30 or 40 years later. In the experiment, middle-aged men who couldn't manage a 10-minute mile—meaning many, many middle-aged men—had about a 30 percent greater risk of developing and dying from heart disease than the small set of fit midlife males who could tick off an impressive 8-minute mile. For

40- and 50-year-old women, the times were a 9-minute mile to achieve the lowest risk of heart disease at age 70 or 80, and a 12-minute mile or slower as an indicator of those who were the most likely later to suffer from or die of heart disease.

Yes, I performed a mile test run after reading this. Go ahead and complete yours now (provided you have no known heart risks or other health problems, of course). And if you land above the at-this-pace-you-will-die-prematurely threshold, do not panic. A properly structured aerobic-exercise program can, by all indications, alter the future.

First, some vocabulary. Aerobic exercise, in contrast to most types of strength training and the short, sharp activity bursts of sprinting, tests "the ability of the circulatory, respiratory, and muscular systems to supply oxygen during sustained physical exercise," according to a recent review article in the *Journal of the American Medical Association*. During sustained cardiorespiratory-centered exercise, the heart and lungs begin to pump and pulse at a much higher rate than when we're sitting. If your resting pulse rate—the number of beats per minute that your heart takes while you're not being active—is in the 70-beats-per-minute range, which is about average for a healthy but not overly fit American adult, it will rise quickly in the first few moments of a cardio session, passing 100 beats per minute and climbing to 140, 150, or higher.

The word *aerobic* means "with oxygen," while *anaerobic*, not surprisingly, means without. Both aerobic and anaerobic exercise create energy through the process of glycolysis, or the conversion of glucose (stored sugar) into fuel. But while aerobic exercise uses oxygen to break down glucose, anaerobic exercise uses other, less efficient but faster mechanisms. Anaerobic glycolysis can't be sustained for long, which is why a hard sprint, during which your muscles contract with too much force and speed to rely on oxygen, leaves you spent. Most of us can't sustain anaerobic activity for more than a few minutes.

But aerobic exercise, in a well-trained body, can continue for hours.

The key, though, is training. All of us start with an innate aerobic capacity, for which we can blame or thank our parents. This is our maximal oxygen capacity, or VO2 max, a measurement that I've mentioned before and that peppers any description of exercise training. A person's VO2 max represents the greatest amount of oxygen that he or she can take into the lungs and distribute to working muscles. Technically, it is the number of milliliters of oxygen that you use in one minute of activity per kilogram of body weight. It's determined, in most cases, by means of an elaborate treadmill test, during which you wear a heart rate monitor and a kind of gas mask that measures your oxygen intake with every stride. During the test, researchers gradually increase the treadmill speed and you, the runner, are forced to exert more energy to avoid falling over. In general, oxygen consumption increases in a linear fashion as physical effort increases; the harder you run or cycle or otherwise aerobically exercise, the more oxygen you draw in and use. But that process continues only up to a point. When you reach some given amount of exertion, which varies from person to person, your oxygen intake plateaus. You can't take in or distribute more oxygen. That is your VO2 max. (It's possible to estimate your VO2 max using the results of your mile test, although the number is a broad approximation only. See the end of the chapter for details.)

A person's VO2 max at any given time is a useful measure of fitness, but despite all of the attention lavished on the number in sports science and by many coaches and athletes, it's ambiguous in practice. Repeated studies of elite athletes, including Olympians, Tour de France riders, World Cup soccer players, marathon winners, national-team rowers, and even British cricketers, have found that the world's best endurance sports participants all have high VO2 maxes. If you don't have a voluminous aerobic capacity, you can't compete. But the athletes with the very highest VO2 maxes are not necessarily those who actually win the Olympic medals or marathon races or cricket batting titles. An athlete with a lower VO2 max than many of his competitors may still set a world record. There's much more to athletic performance than VO2 max.

That being said, most of us could stand to improve our cardio-respiratory fitness, as represented by VO2 max, whether we're recreational, nonelite athletes, health-minded book readers, or those whose mile times were just a bit worrisome (who, in all honesty, are probably the same people). I know mine needs to come down.

It's Good to Be a Beginner

The world record for the mile, at press time, is 3:43.13, or 3 minutes and a shade over 43 seconds, a mind-bogglingly fast record set by the Moroccan runner Hicham el-Guerrouj in 1999. It came 147 years after the first officially timed and recognized competitive mile, which was won in a time of 4:28, or 4 minutes and 28 seconds, during a race in England in 1852. Records before that time were not considered official, because track distances weren't standardized. Once officials began to measure and certify the length of tracks, records could be kept and compared. In 147 years, the men's world-record mile time dropped 45 seconds, an overall improvement of 17 percent.

There are several noteworthy aspects to that progression. First, most of it occurred after the late 1920s, after Paavo Nurmi, the world-record-collecting Flying Finn, and his coach introduced nuanced, scientific training methods, including interval-style sessions, to running. And second, the 17 percent improvement, clawed out over a century and a half of hard racing, is not close to the amount of betterment in fitness and performance that a typical novice runner, cyclist, or other entry-level endurance athlete will see in her or his first year of training.

"The nice thing about being a beginning runner" or other endurance athlete "is that beginners see very big gains in performance," Joe Vigil, a legendary coach who's worked with many of America's best marathoners and other distance runners, told me. "It's one of the things that inspires beginners to stay with the program, I think."

Just how much improvement a beginning exerciser can expect in fitness and athletic performance in the early stages of a new cardiovascular exercise routine varies, but the anecdotal evidence, including estimates from coaches and practitioners, would suggest that a reliable, mathematical estimate would be "lots."

"Someone who starts walking almost every day, if they haven't been walking for exercise before, will probably see their resting heart rate drop by at least a few beats per minute quite quickly," says Dr. Michael Joyner, a researcher at the Mayo Clinic in Minnesota and an expert on exercise physiology. If a walker bumps up the intensity of her or his exercise sessions to a jog, "their resting heart rate may drop into the sixties from the seventies" within the first twelve months.

Such changes occur because the body remodels itself rapidly and dramatically in response to aerobic exercise. The heart, like any muscle being asked to exercise, grows stronger. Studies with sedentary mice given access to running wheels show that cardiac muscle cells begin to lengthen within days after the mice start running, improving cardiac function. Gene markers of molecular activity within individual cardiac cells also increase in mice that take up running.

In people who exercise regularly, the heart often grows demonstrably. In this condition, known almost poetically as "athlete's heart," the chambers of the heart enlarge. The pumped-up heart pushes more blood into the arteries with each stroke, allowing it, most of the time, to pump less often. A trained athlete can have a resting heart rate in the range of forty beats per minute, which in a sedentary person would signal pathology. "Doctors sometimes worry when they see" an athlete with such a low resting heart rate, says Dr. Paul Thompson, a widely acknowledged expert on athletes' hearts, as well as, himself, a marathon runner. "But it's a normal, healthy physiological adaptation to aerobic exercise. It's how a heart is supposed to look and perform. It's just that most people aren't fit enough" to develop athlete's heart.

At the same time, the blood vessels are becoming better able to handle the demands of this muscular heart by increasing their pliability. Studies have found that cells in the walls of blood vessels proliferate after people begin exercising, allowing the tubes to stretch and remain flexible even as extra amounts of blood—and attendant oxygen—flow through them. Working muscles want more blood during exercise. Trained vessels better comply. Afterward, at rest, these vessels remain flexible, resulting in lower blood pressure. Over time, regular exercise also results in the creation of more capillaries (small blood vessels) that run from the arteries to the muscles, easing blood flow during exertion.

The lungs change, too, as a result of endurance exercise. The increased force of every stroke from the heart drives more blood to the lungs, which causes more blood to flow to the air sacs, leading to an increase in the amount of air that is drawn in. The lungs also become more responsive to messages from the respiratory centers of the brain telling them to suck up more air. And the respiratory muscles in the abdomen, which help to push air into and out of the lungs, grow stronger as they are used more. Exercise, which demands more oxygen, creates the conditions under which your body can draw in and distribute more of the stuff.

All of this means that through exercise you raise your VO2 max, typically within weeks of starting a regular endurance exercise program. In some studies, novice exercisers have increased their VO2 max by 30 percent or more after starting cardio. Sometimes the volunteers have jogged on a treadmill. Other times they've ridden stationary bicycles or begun swimming or, in a couple of interesting recent studies of older adults, taken up tai chi. But in every instance, their aerobic capacity has grown significantly and quickly—particularly, says Dr. Thompson, "if they lose weight," since VO2 max is determined partly in comparison to your body weight.

Then a year or so passes, they stop being beginners, "and the improvements become more incremental," Vigil says. For people who want primarily to achieve and maintain a base of good health and fitness, that's probably fine. You can continue the same exercise program for life. But if your ambitions have broadened—and many people who start jogging or bicycling for fun and fitness discover an unexpected, lusty competitiveness in themselves—then it's time to change how you work out. It's time to pull up your pants, buck up your resolve, and train.

Fartlek!

In the 1930s, Sweden was being walloped in international distance running by neighbor and rival Finland, and Sweden resented it. Not only was Finland's marvelously talented Paavo Nurmi collecting Olympic medals and worldwide acclaim, but other, lesser members of the Finnish distance-running squad were trouncing the Swedes in regional and European track meets, the blond leading the blond, and causing consternation back in Stockholm. So the Swedish men's national team coach, Gosta Holmer, a one-time bronze medalist in the Olympic decathlon, decided to imitate the Finns. Most of them at that time were following the brutal example of Nurmi and driving themselves through punishing if short practices once a week or so.

Slyly, Holmer instituted the same kind of regimen among his runners, but instead of confining them to a track, he sent them out into the hills and fields of Sweden and had them sprint toward far-off trees or rocks, then run at a slow speed for a while, then sprint again to the next towering fir. He told them that these workouts were "speed play," or in Swedish, "fartlek." And with that, he changed the nomenclature of running and, to a significant degree, the nature of scientific training.

There are many ways to structure a training program to improve your endurance, and there's surprisingly little scientific backing for any of them. The "ideal distribution" of training time between long, slow workouts, hard, short ones, and something in between "is not well-established from a scientific standpoint," according to a recent review of the science about training.

There are, however, certain elements that almost everyone, from coaches to physiologists to the laziest of competitors (i.e. me), accepts must be part of any cardio exercise program designed to make you faster, as well as fitter. The first and most obvious, but often overlooked, aspect, says Joe Vigil, is "volume." If you want to be a better runner, swimmer, or tai chi-ist, you must engage in ever-increasing hours of running, swimming, or tai chi-ing. The body, through simple, brute practice, becomes accustomed to the motions of running, swimming, or a sui boo pose. You develop neuromuscular "grooves." Several studies of novice marathon runners have shown that one of the most reliable indicators of how well a first-timer will perform in a marathon is the number of miles that he or she ran in the month prior to the race. Those who ran sixty miles per week or more performed significantly better than those who ran forty miles per week or less.

Of course, if you're currently running fifteen miles a week, those distances may sound daunting. But adding volume to your training is the easiest change you can make. No new expertise is needed. You just do a little more of exactly what you've been doing. A standard rule of thumb says that you can increase your training volume by 10 percent a week. If you're running ten miles a week now, next week you can run eleven. Most experts say that adding more than 10 percent a week to your mileage or exercise hours

courts overuse injuries and fatigue. But there is little scientific support for that idea. If you want to add 15 percent to your training
volume, try it for a week. If your legs or body ache unduly afterward, especially if your muscles are still sore and tired after seventy-
two hours, back off and decrease your training volume.

Then look to your speed. When a group of experienced but not
professional runners were recently assigned to a training regimen
that emphasized either additional mileage or reducing the runners'
overall training volume and adding two short but intense interval
sessions each week, the group performing the intervals showed
greater improvement in their subsequent 5K race performances.
"For athletes who are already trained, improvements in endurance
performance can best be achieved through high-intensity interval
training," the study's authors concluded.

However, as this book's author has previously mentioned, intervals can be unpleasant. To be effective, they must use a painfully
large percentage of your available aerobic capacity. The "ideal intensity of exercise" for interval training sessions, a recent study of competitive cyclists determined, is "78 to 93 percent of VO2 max."

But nailing down at just what percentage of your aerobic capacity you're operating is difficult and getting actually more difficult as
scientists study the process more closely. For years, most of us (or at
least those of us who want to compete and/or improve performance)
were told to perform intervals at about 75 to 90 percent of our
maximum heart rate, a number that was believed to be more acces-
sible and useful than our VO2 max. Every gym in the country
contained a poster showing heart rate percentages and target-heart-
rate training zones. We were directed to determine our maximum
heart rate by the simple expedient of subtracting our age from 220.
Using that formula, every 40-year-old man or woman's maximum
heart rate would be 180 beats per minute, and 85 percent of that, a
good range for performing intense intervals, would be 153 beats per
minute.

Many of us, including me, invested in pricey heart rate monitors.

But recently, scientists reran the heart rate numbers. The original charts had been developed in the 1970s for use by cardiologists dealing with ill patients. When researchers in this decade performed new maximum heart rate tests on healthy people, male and female, the formula turned out to be too simplistic. For women, especially, the standard estimates of maximum heart rates proved to be much too high. The formula of 220 minus a person's age overestimated a typical woman's heart rate by about 8 beats per minute, according to researchers at Northwestern University, in Evanston, Illinois, who studied more than 5,500 women aged 35 to 93. The correct formula, the researchers determined, should be 206 minus 88 percent of a woman's age. This construction doesn't trip easily to mind or fit neatly on posters. But it is more accurate. Using these numbers, a typical woman who wanted to exercise at 85 percent of her maximum heart rate should allow her pulse to rise to only 145 beats per minute, not 153.

But even that number is questionable if the woman is past age 40, overweight, or genetically prone to a bird-fast or more sluggish pulse rate, other recent studies have shown. The standard formula also is inaccurate for men once they're past age forty-five, a small study from Norway found. In fact, multiple studies of experienced runners and cyclists have shown that heart rate is, in general, an iffy indicator of true physiological effort. A representative recent experiment involving cyclists who wore heart rate monitors and also underwent exacting physiological testing of their effort and fitness showed that the heart rate numbers provided by the monitors underestimated each rider's true oxygen usage while pedaling by about 6 percent and overestimated his total energy usage by more than 13 percent.

But if heart rate monitors are untrustworthy, then what can you rely on to tell you how hard you're working? As it turns out, the best recent science suggests, you can rely on you. According to a large body of experiments, a person's rating of perceived exertion, or RPE, is a better indicator of actual physiological effort than standard heart-rate-based formulas. The formal RPE tables used by scientists in many experiments employ a scale of 6 to 15—not 0 to 10, as you might expect—since the numbers are meant to correspond, at least glancingly, with pulse rate, which can be measured as a multiple of 6. With this scale, 6 represents immobility and 20 is maximal, unsustainable effort. Activity that rates a 9 would be an easy walk, according to measurements from the CDC. Your individual intervals would need to be conducted at an RPE of about 15, which you'd need to sustain for several minutes or more, most studies show. You can also, of course, use a 0–10 scale, which most of us would find more logical. By that measure, you'd need to make yourself work at an RPE level of about 6 or more for several minutes.

To put all of these abstract and admittedly rather intimidating numbers into practice, try a fartlek. Better yet, announce to friends and colleagues that such is your intention. A fartlek workout is, in its essence, an unstructured interval training session. You can fartlek anywhere. Just start out with ten or fifteen minutes of easy jogging (or cycling or swimming, although most people who fartlek are runners). Then sight a landmark farther along your course. It can be a tree, the top of a short hill, or a crack in the sidewalk. The only requirement is that it be far enough away or high enough that you'll need to sprint for several minutes to reach it. Then sprint. Make yourself work. Your sense of the effort involved should hover at least at a 5 (on the 0–10 scale of RPE). After you reach the object or scale the hill, slow down. Jog or run lightly until your heart rate has stabilized and your RPE has dipped to, say, a 3. Then find another tree and *lek* with *fart*.

What you are aiming to dismantle through such workouts, whether you realize it or not, is the tyranny of fatigue.

The Swish Test

Not long ago, at the Institute for Biomedical Research into Human Movement and Health, at England's University of Birmingham, eight male, highly fit, competitive cyclists pedaled furiously on eight cycling ergometers (computerized stationary bicycles). Machines measured their heart rates, pedal cadence, and pedaling power output. Researchers circulated between the panting riders, offering them fluids.

The fluids weren't for drinking. They were for messing with the riders' minds. Each cyclist took a long sip, swirled the fluid in his mouth for ten seconds, and spat it out into a bowl conveniently held for him by a scientist. No one swallowed. Everyone kept pedaling.

Some of the fluid was disguised water, heavily flavored with an orange sweetener. It contained no calories, no carbohydrates, only a sugar-free Tang-like flavoring. Other water bottles held fluids laced with a hefty dose of glucose (liquid sugar), in addition to that same orange flavoring. The final type of fluid contained maltodextrin, a flavorless carbohydrate, plus the sugar-free, orange flavor. The fluids tasted the same. None of the riders knew which fluid they were receiving.

No one, remember, swallowed. No one ingested any actual calories or fuel.

But the riders who had swished and spit fluids laced with either glucose or maltodextrin, which are carbohydrates, the body's preferred fuel during strenuous exercise, finished their time trials significantly faster than the riders who'd spit only water. Their heart rates and power output also were substantially higher, meaning that they had exercised demonstrably harder than the riders who'd spit water. But when researchers asked the cyclists how they'd felt during the ride, those who'd rinsed their mouths with carbohydrates shrugged and said, in essence, that while any time trial is hideously unpleasant, this one, by that standard, hadn't been too bad. The cyclists who'd rinsed with water said they felt exhausted.

What causes the aerobically exercised body to grow tired is a matter of surprising scientific contention. Until recently, most researchers would have said that thinking and mental processes played little role. Muscles failed, physiologists thought, because of biochemical reactions within the muscles themselves. They began getting too little oxygen and became doused with too much toxic lactic acid. They stiffened and seized.

But that theory began to unravel some years ago, when a number of researchers independently began to question the role of lactic acid in particular. The researchers could find no signs that lactic acid buildup affected the ability of muscles to contract. Instead, they found persuasive evidence that lactic acid, which is created by muscles during glycolysis, is in fact a fuel. The mitochondria within muscles burn it as a backup energy source. As the title of a recent review article in *The Journal of Physiology* announced, "Lactate Is Not Evil."

There are other problems with the idea that fatigue involves only the muscles. "We know that people speed up at the end of exercise," says Ross Tucker, Ph. D., a researcher with the Sports Science Institute of South Africa, who studies fatigue in athletes. If calcium or other biochemical changes in the muscles "caused muscle failure, this would be impossible at the end, when these changes are at their greatest levels."

A remarkable experiment recently completed in England found that if they were lied to, people could in fact race harder than they thought they could. For the study, a group of recreational bike riders completed a time trial during which they were told to ride as hard as they possibly could through four thousand meters. Then they were told that they would be racing against an avatar version of themselves, a computer program set to re-create each rider's best effort. In one race, that's what they did: The avatar exactly matched the rider's original time trial pace, and in response, the riders finished in the same time as they had achieved in their first ride. They told the scientists that this race against themselves had utterly exhausted them.

But in a subsequent trial, the avatar program was subtly changed, without the volunteers being told, so that the program pace was now 2 percent faster than the rider's own original—and supposedly fastest—speed. Faced with this new program, which they believed to be their own prior best ride, the cyclists almost invariably pulled something from within themselves and rode harder than they ever had. They finished their rides at a pace 2 to 3 percent faster than they had before. Somehow, their brains and bodies, facing a reality that they didn't know was skewed, altered the moment at which fatigue set in.

Dr. Tucker does not find such studies surprising. He and many other (but not all) physiologists now believe that exhaustion doesn't involve only changes in the muscles; it also involves the brain. "What we now think is that the muscle isn't acting on its own," he says. "There's an interplay of central processing and muscular exertion."

From the outset of endurance exercise, in fact, "the brain asks for and gets constant feedback from the muscles and other systems" and checks on "how things are going," says Carl Foster, Ph.D., a professor in the department of exercise and sports science at the University of Wisconsin, La Crosse. Through mechanisms that aren't fully understood, the brain tracks and calibrates the amount of fuel that is in the muscles, as well as the body's core temperature. As the amount of fuel drops and the temperature rises, the brain decides that some danger zone is being approached. In theory, after all, prolonged endurance exercise by humans could have all kinds of hideous physical consequences. If there were no brakes in place, strenuous exercise would almost always result in heatstroke, or severe, relentless muscular contractions would snap bones. This happens sometimes in racehorses. It never occurs in people, because before that degree of effort is reached, the brain starts reducing "the firing frequency of motor neurons to the exercising muscle, leading to a loss of force production," says Ed Chambers, Ph.D., a researcher at the School of Sport and Exercise Sciences, at the University of Birmingham, and an author of the carbohydrate-drinks study. In other words, the mind, recognizing that the body may be going too hard, starts sending fewer of the messages that tell the muscles to contract. The muscles contract less frequently and more feebly. In a sensation familiar to anyone who exercises, your legs die beneath you.

The mental choreography of fatigue is intricate, though, involving messages sent not only from the brain to the straining muscles but also from one part of the brain to other parts of the brain. Data from some recent brainwave studies of athletes showed that during long, hard exercise, there's often a moment when portions of the brain become "de-aroused," Dr. Foster says. "It's similar to depression," he adds, and it plays out in motivation. You begin to wonder why in the world you're running, swimming, or pedaling so hard. You slow down.

It's the ability that interval training seems to have to recalibrate, at least slightly, the brain's sense of how much you can handle that makes those workouts so potent, Dr. Tucker says. If fatigue occurs not simply because muscles get tired, but because the brain tells them they're tired (even though they almost certainly have some reserve fuel and strength remaining), then you can teach your brain to hold off a bit longer, if you should so desire. "I think the training effect of the theory that fatigue occurs in the mind as well as in the muscles is potentially very profound," Dr. Tucker says. In that case, training "is no longer simply an act of getting the muscles used to lactate or teaching the lungs how to breathe harder." It's also about getting your brain to accept new limits by pushing yourself, safely.

You could, of course, try lying to yourself, although it's difficult to see quite how. And having someone else yell deceptive appraisals of your performance at you won't help much, either. Studies in which researchers have given volunteers "deceptive feedback," such as telling them that they're running slower than they actually are, typically result in the athletes slowing down, not speeding up. Deception in these instances is "demotivating," the authors of the studies in question wrote. So you're probably best off, Dr. Tucker says, continually pushing yourself through interval sessions. Try, at some point, to reach an RPE of 10 on the 1–10 scale. Then see if, on another day, you can, in the immortal advice from Spinal Tap, "put it up to eleven."

"Once your brain recognizes that you're not going to damage yourself," Dr. Foster says, "it'll be happy to let you go."

The Rest

After the workout you must rest. Aerobically exercised bodies clamor for repose. Ask almost any elite marathoner what she does in the hours between workouts and she'll tell you, "nap" (and eat and then possibly nap again). "If you don't rest, your body can't consolidate its gains," Dr. Joyner says.

Whether this rest has to be absolute or can be achieved in the form of so-called active recovery is undetermined, from a scientific standpoint. Studies to date have been conflicting, with some finding that a day or even two of physical inactivity every week produces better race times in subsequent weeks than "rest days" that involve light exercise. Other studies, though, have found that activities such as yoga, which in most incarnations is not aerobically demanding, lessen the inflammation and lingering muscular damage that follow repeated cardio workouts. One recent study of yoga practices in India found that classes soothed the muscles and cardiovascular systems of athletes enough that in subsequent training sessions the athletes could exercise for longer before feeling exhausted. Another study, this one conducted at the Ohio State University College of Medicine, found that while hatha yoga classes didn't result in demonstrably lower inflammation levels in people who'd been exercising or undergoing other types of stress, they did leave the participants feeling more relaxed and energized.

In other words, yoga, based on the limited, current science, represents a fine compromise between inactive and active rest. This is good news, since I had planned to continue attending yoga classes whether scientists approved or not.

Meanwhile, science related to another popular type of athletic rest, tapering, is sketchy Tapering, or a multiday layoff from exercise before a competitive event, is "poorly understood," a recent review concluded. But the review, which looked at all the most recent studies, added that the weight of the evidence suggests tapering is useful if you compete. A compelling South African study of bike racers found that those who reduced the intensity of their workouts by 50 percent in the two weeks before a time trial finished more speedily than those who hadn't tapered. Other studies have found that recreational runners who drop the volume of their training by 30 to 75 percent improve their 5K times by as much as 6 percent.

The various studies unhelpfully do not clarify, however, whether reductions in mileage of 30, 50, or 75 percent are the most efficacious. So you're on your own. In general, the science suggests, you should maintain some intensity in your workouts but drop the volume in the two weeks prior to an event.

Finally, monitor yourself for overtraining. It's astonishing how often those of us who love endurance exercise flame out as a result. Overtraining is particularly common, in fact, among people who are the most dedicated to their training. (I've never overtrained.) By some estimates, 60 percent of endurance athletes will overtrain at some point. "I work with a lot of runners and distance athletes," Ralph Reiff, the director of St. Vincent Sports Performance, in Indianapolis, and an expert on overtraining, told me. "In my experience, a large percentage of the people who train for ten-Ks, half marathons, and marathons are overtrained by the time they reach the starting line. Same with cyclists and cross-country skiers. A very high percentage get into a state of fatigue that they just cannot get out of."

The most frustrating thing about overtraining is that it's an outgrowth of proper training. "You can avoid overtraining by under-training," says Bob Larsen, an elite running coach, who works with many of American's best distance runners, "but then you won't win."

"You expect a peak," Reiff agrees, "but you get a slump." Why? Although scientists don't really know what causes overtraining, it's likely that constant training leads to an overreaction by the body's immune system. It starts producing too many inflammatory molecules, which communicate with circulating blood cells and cause them to ramp up the production of chemicals that can inflame the whole body. This theory (which is hard to test, since ethical review boards frown on experiments that induce overtraining in people) may explain the wide range of symptoms often seen in overtrained athletes, including mood changes, apathy, insomnia, and exhaustion. Genes are also involved. Ongoing research funded by USA Track and Field suggests that strenuous cardio training alters the ways in which a wide array of genes work, and most of the alterations improve bodily function. But if some genes become either exceptionally active or suddenly quiescent under the strains of training, you can wind up with a body that responds to exercise inappropriately, one that is overtrained. Someday, blood tests may detect such changes early and help most of us avert overtraining.

But for now, monitor yourself. If your training and race times are rising and your motivation and energy levels are dropping, back down. "There's only one palliative for overtraining that I know of," says Robert Schoene, M.D., a professor of medicine at the University of California, San Diego, who has written about overtraining and has treated afflicted athletes. "Rest, rest, and more rest."

Which is the last thing that many of us wish to hear. "The problem with your typical overtrained athlete is that he thinks if he just pushes harder, he'll be better," Larsen says. "But that's how he got overtrained in the first place. If he were lazy, this wouldn't have happened."

So both as a preventative and, if necessary, a cure, be sure to incorporate into your training rest, rest, and more rest. "The human body, no matter how strong and fit it is, must have downtime," Dr. Schoene says.

The marathoner Alberto Salazar is famous among runners for having permanently compromised his health and shortened his career by never laying off. Eventually he suffered a heart attack in his forties. "Alberto gets used as a worst-case example, to scare runners into taking it easy sometimes," Larsen says.

As for me, I've taken the Salazar example to heart. I still run. In all the decades since college, I've never quit. For many years, I raced, finishing 5Ks and 10Ks and a few lacerating, exhilarating marathons. My second date with my husband was at a 10K race along the lakefront in Chicago. I beat him. I've made friendships through running. I suspect that I've kept my figure, to the extent that I have, through running. I've also taken up mountain biking, which I love, and road-bike racing, which I didn't (too much jostling at high speed, in a tightly packed scrum of riders, wheels inches apart, knowing that any mistake by anyone will send you all to the hospital; not a sport for pansies, whom I'd direct to something gentler, like football).

But always I've run. I just don't run as much these days. I have two dogs, a border collie and a shepherd, who, like their owner, are getting older, and they serve as patient, willing training partners. Together, we jog companionably for three or four miles, with no concern for heart rate or RPE or personal best times. The sky arches overhead. My heart thrums. My dogs settle in at my pace. We're not fast anymore, but we are out there four or five times a week, summer and winter, sunshine and sleet, running. That's what I mean when I talk about endurance.

How to Test, Improve, and Tunefully Enrich Your Endurance Training

1. Run a Mile in Your Shoes.

Clear the activity with your doctor first, of course, but then recruit a buddy or use a timing watch to find your mile-run time. This can provide a good measure of your current aerobic fitness and also serve as a window into the future. A man in his 40s who can run an 8-minute mile is in the highest fitness category, according to research from the Cooper Institute. (That's a 7-minute mile if you're age 30 or younger.) A woman of the same age who can run a 9-minute mile is also quite fit. Middle-aged men who can't run a 10-minute mile or women whose time is 12 minutes or higher are the least fit and may have a higher heart disease risk in later life. The good news: You can improve your mile-run time, whatever your age.

2. Do the Math.

Scientists at Brigham Young University recently published a formula—complicated enough to make a person's eyes cross—for determining VO2 max based on your mile-run finishing time. Despite the many parentheses in this equation, it is only an approximation of VO2 max. But if you'd like a fair estimate of your current aerobic capacity, the formula is: VO2 max = 100.5 − 0.1636 (your body weight in kilograms) − 1.438 (your mile time) − 0.1928 (your resting heart rate) +

8.344 (1 if you're a man; 0 if you're a woman). For comparison purposes, a typical VO2 max for a sedentary middle-age man is about 35 and for a woman it is about 30. Well-trained athletes can have a VO2 max of 80 or higher in men and 60 or higher in women.

3. Reset Your Treadmill, If You're So Inclined.

The body doesn't discriminate between different types of endurance exercise. You can run, cycle, swim, or cobble together any combination of these and other activities if your goal is to improve cardiovascular fitness. But in practical terms, running on a treadmill is not the same as running outside. You face no wind and no terrain changes indoors. In one of the few direct comparisons of treadmill and overground running, scientists found that sessions on a flat treadmill required about 5 percent less energy than running outside. To better approximate running over ground, the scientists concluded, you should set your treadmill to a 1 percent grade.

4. Add Speed. Then Drop It.

Strong science about how best to structure a training program is woefully slight, but what studies do exist suggest that if you plan to race, you need to incorporate speed training—aka high-intensity intervals—into your routine. You can do this formally by performing a certain percentage of your workouts at a given percentage of your VO2 max. One recent study of elite cyclists concluded that those numbers should be, ideally, 15 percent of your training at 85 percent of your VO2 max. But most of us don't need to be so systematic. Fartlek, or speed-play training, involves simply incorporating long sprints into your regular runs or other training sessions. This makes the workouts hard but, for the most part, fun. Don't forget, though, to rest after strenuous workouts. And tapering, or cutting back significantly, your training volume in the two weeks before a race has been found, in a few studies, to improve subsequent race performance.

5. Keep a Training Diary.

Studies have shown that logging your mileage every day, while also noting your times if you were performing speed training and how you felt (tired, discouraged, pleased, etc.), provides valuable feedback about your progress. It can also sound early warnings about overtraining. If you notice that your times are going up and your motivation is falling, try reducing your training and see if you start to feel and perform better.

6. Leave It to Bieber.

As consolation for suggesting that you add intervals to your training, I will now point out that the right beat may make endurance training easier. In a fascinating recent study, British researchers asked twelve healthy male college students to ride stationary bicycles while listening to music that, as the researchers primly wrote, "reflected current popular taste among the undergraduate population." Each of the six songs differed somewhat in tempo. The volunteers were told to ride the bicycles at a pace that they comfortably could maintain for thirty minutes. Then each rode in three separate trials. During one, the six songs ran at their normal tempo. During the other rides, the music's tempo was slowed or increased by 10 percent. The riders didn't know this was happening. But their riding changed in response. When the tempo slowed, so did their pedaling and their entire affect. Their heart rates fell. Their mileage dropped. On the other hand, when the tempo of the songs was upped 10 percent, the men covered more miles in the same period of time, produced more power with each pedal stroke, and increased their pedal cadences. They knew that they were working harder; their ratings of perceived exertion rose. But they didn't care. When the music was played faster, the scientists wrote, "the participants chose to accept, and even prefer, a greater degree of effort." So before your next run or bike ride, crank up the tempo with some insidiously catchy Lady Gaga download (or Justin Bieber or Katy Perry or whatever reflects the current popular taste in your household), and load it on your iPod. "Our bodies," says Nina Kraus, Ph.D., a professor of neurobiology at Northwestern University, who studies the effects of music on the nervous system, "are made to be moved by music and move to it."

6

Give Me the Strength to Carry On

Here is a simple illustration of the importance of strength. A few years ago, scientists at the Boston University School of Medicine created a genetically altered mouse that had what was essentially a push-ups gene. When the gene was activated, the animal added what are known as Type II muscle fibers, very much like people do when they lift weights moderately or do push-ups. Endurance exercise adds different muscle fibers, while heavy-duty weight lifting adds more muscle volume overall. This was not a model of serious weight lifting. The genetically altered animals did not become bulky little muscle-bound Schwarzenegger mice. Instead, the gene that they carried prompted their existing muscles to grow somewhat thicker and firmer, so that their hind- and forelimbs resembled Michelle Obama's upper arms. Sleek and lean, they moved about their cages with ease and grace. Even when they were placed on a diet of high-fat chow, they gained little flab, remaining healthy and sinuous.

When the scientists used other biochemical processes to switch off the gene, however, the mice were soon in trouble. They lost muscle fibers, reducing their muscles' size, but worse, they gained fat, growing obese with dizzying speed. They also developed insulin resistance—a precursor to diabetes—and a marbling of fat throughout their livers, a condition that contributes to multiple health problems. Too pudgy for their now spindly legs to support them, they plopped onto the floor of their cages and stayed put.

But reactivation of their push-ups gene returned them soon to shiny, happy rodent-hood, even if they remained on a high-fat diet. As their muscles regained strength and vigor, their metabolisms and other physiological systems did as well. They burned off the fat that had been stored throughout their bodies and lost their insulin resistance and much of their liver fat. They began moving about their cages more. They were once again good-looking, swaggering, lean little animals.

The results show, the authors concluded, that the "growth of muscle can regress obesity and resolve metabolic disorders." Strong muscles, in other words, confer many physiological rewards, and those benefits flow from a process that, in human terms, would require only performing a few push-ups or perhaps a medicine ball toss or two.

Getting Past "The Aerobics Way"

Who needs to be strong? That question isn't as easy to answer as you'd think. There was a time, and I'm not sure it's ended, when many fitness experts, coaches, athletes, and everyday exercisers thought that strength training was unnecessary and undesirable for most of us.

Certainly, Kenneth Cooper, the physician who helped to launch the fitness boom back in the 1970s and 1980s, was dismissive of strength training. In bluntly titled books such as *Aerobics*, *The Aerobics Way*, *The New Aerobics*, and *Run for Your Life: Aerobic Condi-*

tioning for Your Heart, he urged people to complete as much endurance or "cardio" exercise as they could, on the assumption that only aerobic exercise would improve heart health, discourage the development of other diseases, and aid in weight control.

"Exercise has been endurance-centric for quite some time," says Stuart Phillips, Ph.D., a professor of exercise science at McMaster University, who studies the effects of strength training.

In fact, strength training was and to a degree still is openly discouraged in some endurance sports circles. Dr. Cooper and other authorities like him have suggested that lifting weights could result in a person's adding extra muscle mass. More mass would make you bulky, which, if you were a runner, cyclist, or swimmer, would slow you down. You'd be propelling more poundage through space. The physics seemed ineluctable.

But those authorities didn't fully understand physiological strength and power and how your muscles actually respond to being pushed.

Violent Contractions

Muscles are among the most complicated tissues in your body and, in certain ways, are the weirdest. Scientists remain divided on whether skeletal muscle is even a tissue or an organ. If an organ— meaning that it contains more than one type of tissue—skeletal muscle would supplant skin as the largest organ in the body and be a killer trivia-question answer.

For starters, muscle cells aren't cell-shaped, in the way that most of us think of cells—they aren't circular. There are three types of muscle: smooth, which lines blood vessels; cardiac, found in the heart; and skeletal, which is most of the muscle in your body. Skeletal muscle cells are long and thready. A single cell can run the full length of a muscle, although most are shorter. Each muscle cell is composed internally of countless sarcomeres, interlocking bricks of different lengths that are themselves made up of linked protein filaments. The filaments are alternately thin and thick, giving skeletal muscle a distinctive striped appearance. The thin protein fibers are known as actin, while the thicker fibers are called myosin. Myosin fibers have tentacle-like offshoots that latch on to the neighboring actin threads.

When a muscle receives a signal from the brain ordering it to contract, the protruding bits of the myosin protein heave like microscopic oarsmen, catapulting the actin protein past them. Then the process reverses, tightening the muscle. Contractions can be quite powerful. If all of the fibers in your quadriceps muscles contracted simultaneously, the force would snap your leg bone. But only a third or fewer of the fibers in a muscle contract at any given time, allowing your body to remain intact.

Some damage will occur within the muscle itself, though, if the contraction is in response to a force with which the muscle is unfamiliar. Heft a barbell, hoist a toddler, push yourself up off of the couch—each of those actions applies force to your muscles and requires them to either contract in answer or collapse in a heap, which would be embarrassing. That's why weight training is also called *resistance training*. The terms are interchangeable, along with *strength training*. (*Weight lifting* is different; it's a sport and involves lifting weights: Whoever lifts the most wins.) Your muscles are working against resistance or force. If the force is familiar—if your muscle has dealt with a similar amount of force before—the cell lengthens and shortens without incident. But if the force is beyond what the muscle is used to, it overlengthens, and the actin and myosin separate from each other. Fibers fray.

And that's okay. At that point, satellite cells, the specialized triage cells in your muscles, rush to the injured site, selflessly fusing themselves to the injured muscle fibers and enabling them to regenerate, becoming thicker and stronger.

That process is known as muscle hypertrophy, and it ends with you having a stronger, firmer, but not necessarily noticeably larger biceps, shoulder, or calf muscle, depending on which limb was working against the force. The fibers are thicker, but the muscle itself is not greatly enlarged. It would take an additional and substantial infusion of hormones, such as testosterone, human growth hormone, or other artificial anabolic steroids, to cause the muscle to engorge. You also would need to continuously increase the forces being applied against that muscle; you'd have to keep lifting increasingly heavier weight. And you would need to be mainlining protein. Those are the methods by which bodybuilders achieve their globular physiques.

The rest of us are not going to look like that.

Mutt and Jeff

There is a famous set of photographs often used in physiology lectures that shows a pair of adult identical twins in the 1940s. One twin became a dedicated distance runner. In the photo of him, he wears a loose white singlet and boxy shorts, and his rangy body seems to be all knees, elbows, and collarbone. He's lean, almost gaunt.

His twin, on the other hand, is spherical. A competitive bodybuilder, he resembles a bag of marbles. His upper arms, chest, and shoulders bulge. His thighs round toward each other. He looks shorter than his twin, because he occupies more horizontal acreage.

The photos are often employed to illustrate the differing effects that endurance and resistance exercise have on the body. They, and much of the science and coaching advice that have followed them, have emphasized that weight training builds bulk. But what those photos really show is the extremes of morphology that occur if—and only if—someone dedicates him- or herself to a single exercise unstintingly and professionally.

"Your everyday runner is not going to become super lean, and your average person who starts weight training is not going to become extremely muscular," Dr. Phillips says.

When researchers in Germany reviewed six decades' worth of studies about weight training by healthy young people, they found that even children as young as five or six grew physically stronger after beginning light-duty resistance training programs. But they added almost zero bulk. Another review, which looked at results among the elderly, came to a similar conclusion. People in their seventies, eighties, and even nineties could gain significant amounts of muscular strength if they lifted light weights two times a week, the review's authors wrote. But, again, the participants did not become visibly buff.

Young men college-age or older seem to be the group that adds muscle most easily, research shows. Competitive bodybuilders are also successful at adding muscle, though their results can owe something to pharmacology (a.k.a. steroids).

But what the rest of us gain from strength training is primarily strength and, as a nice sop to the megalomaniacs among us, power, and plenty of it. Maniacal laugh optional.

Words of Power

In an edifying if simple experiment, a group of dedicated recreational runners were taught to weight train. Until then, the volunteers had just been running to prepare themselves for local races. The study took place in Finland, but it has universality. Its subjects were typical of middle-of-the-pack 5K or 10K racers anywhere.

The runners were divided into three groups. One continued to train as they had, with running alone, although each volunteer's mileage steadily was increased. Another group was set up in the gym with trainers and shown how to traditionally lift weights. They completed a routine consisting of upper- and lower-body exercises with machines and free weights, the resistance or amount of weight that they were lifting growing progressively. And the last group underwent explosive power training.

In physiological and practical terms, strength and power are not precisely the same. Strength means how much force your muscles can produce. Power is the ability to channel that force quickly. If strength is about moving weight through space, power is about moving that weight fast, even explosively.

Traditional strength training, using weight machines or barbells and other free weights (so-called because they are not tethered to machines), builds mostly muscular strength, although it can increase power somewhat. In the Finnish study, the runners who practiced traditional weight lifting for eight weeks improved their leg strength and, presumably as a result, their maximal running speed during a timed sprint. They also performed better overall during a treadmill run than the group that had not strength trained.

But the third group, which had practiced explosive power training by jumping onto and off of boxes and lifting light free weights over and over at a faster pace than the traditional strength trainers, also outperformed the running-only group on the treadmill, and had a notable kick during the sprint testing.

Strength training and power training, the authors concluded, were "effective in improving treadmill running performance."

Multiple other studies have produced similar results. There seem to be few if any downsides to weight training for endurance athletes and many advantages, chiefly that the training seems to make people faster. In another representative study of traditional, machine-based weight training involving both the upper body—gangling in many runners—and the legs, a group of experienced runners became physiologically more efficient. They used less oxygen to cover the same mileage as they had before they began resistance work. And their late-stage sprinting ability soared.

Other studies have found particular benefits for endurance athletes, such as runners, from more idiosyncratic explosive-power training, namely plyometrics. With plyometrics, you leap from the ground onto boxes and otherwise overload your muscles using body weight and sudden acceleration and deceleration in space. In a recent study involving competitive male distance runners, half added a weekly session of standard machine-based strength training to their ongoing running training. Another added a once-a-week session of plyometrics. After eight weeks, both groups had improved their running efficiency, but the runners practicing plyometrics had improved more.

Several small studies since of both male and female runners, experienced and novice, have produced similar results. In general, plyometrics "improved strength and neuromuscular performance and enhanced VO2 max and running efficiency" better than lat pull-downs and leg extensions—although those standard weight training exercises provided improvements in running performance in most cases, too.

The impacts of combining strength and power training with your usual running, bicycling, or other endurance training routine may be greatest deep beneath the body's surface. In one of the first studies to look at the molecular effects of adding strength training to a cardio-only routine, cyclists who ended an hour-long bike ride with a series of leg-press resistance exercises had far more genetic remodeling within their muscles than cyclists who did no strength training. Their muscles contained twice as many various signaling molecules that jump-start adaptive changes and make muscles better able to use oxygen—to have, in other words, greater endurance. Resistance exercise, the authors wrote, "amplifies the adaptive signaling response" in the muscles. It redoubles the benefits of the cycling or running.

It also, as other science shows, tunes up an out-of-shape nervous system.

Pump Up the Neuron

An interesting recent study of what allows some cyclists to generate higher watts while pedaling than others found that it wasn't the strength of the rider's individual leg muscles that mattered. A cyclist might have massive quad strength but still be outsprinted by a competitor with spindlier thighs. By attaching electrodes above each of the main muscles in the cyclists' legs and watching how the muscles lit up electrically during pedal strokes, the researchers determined that it was the coordination between muscles that determined a rider's power limits. The better his muscles worked together, the more powerful his cycling could be.

Strength training seems to be a key to improving inter-muscle coordination. "Long before you see" any added muscle mass from a weight training program, says Avery Faigenbaum, Ph.D., a professor of exercise science at the College of New Jersey who's studied the effects of resistance training in children, "you are probably getting neuromuscular improvements." The nerves attached to muscle cells begin accepting and sending signals between the muscles and the brain more quickly. The connections between nerves strengthen. The moving body begins to click along more efficiently.

A few small studies have shown that people who gain almost no muscle mass while weight training, such as children and the elderly, do increase the activation of motor units within their muscles. A motor unit consists of a single nerve cell and all of the muscle cells that it controls. When more motor units fire, a muscle contracts more efficiently. So, in essence, strength training liberates the innate strength of the muscle in the young and the old, activating power that had been lying in wait, unused.

Something even more striking may be happening in adults' muscles. A telling recent study looked microscopically at swatches of muscle tissue before, after, and while fit athletes were in the midst of a program of strenuous weight training. Skeletal muscle cells, unlike many types of cells, can have multiple nuclei, or control centers of the cells. The microscopy showed that, even early in the program, the cells from strength-trained muscles contained evidence of newly formed nuclei, as well as additional nervous system connections. Well before the athletes displayed visible, external evidence of bigger biceps or quads, their muscle cells were remodeling themselves and becoming more intricately woven into the nervous system.

Such effects have been found among people lifting very light weights. In results from studies conducted at the Brain Research Centre, at the University of British Columbia, women in their sixties and seventies who completed an easy resistance training regimen twice a week for six months performed significantly better on tests of strength, physical coordination, and even cognitive function than other comparably aged and fit women who'd joined a stretching and toning class for half a year. Although the researchers didn't conduct muscle biopsies or other invasive procedures on their volunteers, they speculated that the resistance training had aided the women in part by stimulating their nervous systems. Resistance training requires an upsurge in brain usage, says Teresa Liu-Ambrose, Ph.D., a professor at the University of British Columbia and a researcher at the Aging, Mobility, and Cognitive Neuroscience Lab, who led the study. "You have to think about proper form and dealing with technique." Nerves and muscles have to learn to work together, perhaps after decades of relative disuse.

The brain benefits from being used, she continues, so that, in a neat circle, resistance training demands and creates nervous system connections and brain circuitry, even in people nearing their ninth decade of life.

A better question, then, may be, Who among us *doesn't* need to be strong?

Why Resist?

In a study that might well have gone viral if more teenage girls subscribed to the *British Medical Journal*, scientists in Denmark discovered that those of us with sturdy, muscular thighs, typically conferred by strength training, live longer than those with stick-figure slender thighs. Contrary to popular belief, thunder thighs are desirable and healthy.

Expanding on that pleasant finding, Dr. Phillips, the physiologist from McMaster University, and his colleagues have found that a regimen of weight training—without any additional endurance exercise—can in fact replicate most of the health benefits generally associated with running, bicycling, swimming, and walking. In a series of experiments over the past several years at his lab in Canada, he found that weight training could "control blood sugar levels and reduce blood lipids [the fats that contribute to cardiac disease] just as effectively as endurance training and better" than most drugs.

Take diabetes, he says. Weight training may be more effective against it than endurance exercise. Muscle is a major consumer of and storage site for blood sugar. When you use and build muscle tissue with weight training, you increase the muscles' demand for glucose. The muscles pull it from the bloodstream, so blood sugar levels don't rise dangerously. In one large trial of weight training as a preventive for diabetes, a group of sedentary people who'd been diagnosed with insulin resistance, often a precursor to the disease, began standard weight training, while a second group remained inactive. After eight weeks, the weight trainers had better blood sugar control, based on various blood markers, than the inactive group. The American Diabetes Association now recommends that anyone with insulin resistance or full-blown diabetes practice strength training at least twice a week.

The practice isn't bad for weight control, either. In several recent studies, a regimen of fairly light weight training by sedentary

men and women led to a significant decrease in waist circumference
and in the volunteers' levels of visceral body fat, or the avoirdupois
that builds up around bodily organs and has been associated with a
higher risk of cardiovascular disease and diabetes. Sedentary women
assigned to a basic, gym-based strength training program three
times a week, for instance, gained less weight over the course of six
years than women who didn't train. Most of the women, admittedly,
didn't stick with the program for the entire six years. But those who
were the most diligent, who showed up at the gym more often and
regularly than other participants, packed on fewer pounds overall
and significantly less belly fat than women who quit the program
early on or never began.

In Dr. Phillips's lab, weight training has been shown even to
affect cardiorespiratory fitness, as measured by changes in VO2
max, or, as you recall, the maximal amount of oxygen that the heart
and lungs can deliver to the muscles. Most of us, including physiologists, probably would have thought that only endurance exercise
training raises a person's VO2 max. But weight training, on its
own—no running between machines to add an endurance aspect to
the workout—improved people's VO2 max, Dr. Phillips says.

"Many people have this notion of a continuum of exercise," he
adds, "with weight training at one end and endurance exercise,
something like marathon training, at the other. They think that
there is little or no overlap" between the physiological effects that
the two types of exercise produce, "and that most health and fitness
benefits come from the endurance side, while all you get from
weight training are big muscles."

Those ideas, he says, are being overthrown.

A Ripe Old Age

Arguably the most profound health impacts of weight training involve its effects on how we live as we age. Sarcopenia, the inexorable loss of muscle mass that begins when we reach our forties, "robs people of independence, of the life that they want to live," Dr. Phillips says. Someone who is frail, who has too little muscular strength to rise from a chair or walk up and down stairs, is someone whose life is confined and no longer his or her own. "Weight training combats sarcopenia," Dr. Phillips says. "It changes the dynamics of aging."

A review of dozens of studies of resistance training among elderly people reached the same conclusion. "Regular resistance exercise is a potent and effective countermeasure for skeletal muscle aging," the authors wrote.

It also builds bone, another tissue that most of us start losing at middle age. Until fairly recently, most scientists had thought that the best way to bulk up your skeleton was to run. The pounding or impact from striking the ground was thought to deform the bone slightly, stretching cells and forcing them to adapt, usually by adding more cells. This, you may notice, is also how muscle responds to exercise. But studies of distance runners, male and female, show that they do not necessarily have stronger bones than non-runners. So a growing number of scientists think that explosiveness and power are what build bone. You need "large forces released in a relatively big burst," says Alexander G. Robling, Ph.D., an associate professor of anatomy and cell biology at Indiana University School of Medicine, who studies bone mechanics. You need strength training, particularly if it includes plyometrics-style, hopping, explosive moves.

Finally and perhaps most surprising, weight training appears to be the best remedy against the insidious loss of joint flexibility with aging that so many of us know and lament. It is, it seems, more effective than stretching. Adults randomly assigned to a five-week weight training routine that targeted their hamstrings, hips, and shoulders wound up with greater range of motion in their knees, ankles, hips, and shoulders than other adults who did not exercise or who completed five weeks of stretching exercises also aimed at the hamstrings, hips, and shoulders. The stretchers did, however, achieve some improvement in their joint flexibility, particularly in the hips and shoulders. They could move those joints through a broader range of motion. They became, of all the volunteers, the most limber. They also now had stronger leg and upper-body muscles than any of the other volunteers.

But, as Dr. Phillips says, "All in all, weight training may be the most important thing you can do for yourself if you want to maintain control of your life as you age." The problem, he added, is that many of us think that weight training, to be effective, must be complicated, mathematical, painful, and hard-core, and that it must involve sit-ups. None of that, based on the latest evidence, is true.

The Core of the Matter

Like most people in the fitness field, Thomas Nesser, Ph.D., a professor of physical education at Indiana State University, once adamantly believed that the firmer the midsection, the better the athlete. A few years ago, though, he actually tested that idea. Recruiting twenty-nine well-sculpted NCAA Division I football players at the university, a football powerhouse, he calculated the athletes' core-section stability by having them perform a series of sadistic tests, such as medicine ball throws and crunches until they collapsed. Then he measured each player's sports-specific strength and agility in twenty- and forty-yard sprints, vertical leaps, shuttle runs, and other tasks.

The results were "really surprising," Dr. Nesser says. In general, the players with the most rock-solid cores were no better at the sports-specific tests than those with feebler middles. "I can't tell you how many times I ran the numbers and checked and rechecked the results," he says. "I couldn't believe it."

To assure himself that the results were valid, he and his colleagues reran the experiment recently with a group of healthy collegians who were not athletes. The young people lunged, twisted, crunched, and held a rigid plank position to measure the hardiness of their back, abdominal, and side muscles, and then completed a battery of physical performance tests, including leaping off the ground while tossing a medicine ball backward over the head and sprinting through a short obstacle course.

Again, the volunteers with the sturdiest cores did not outshine the others. There was little correlation at all between robust core muscles and athleticism. Despite the emphasis that many coaches, trainers, and athletes themselves place on "core training for increased performance," Dr. Nesser and his coauthors wrote, "our results suggest otherwise"—and raise reverberant questions about just what the purpose of strength should be.

Ask many coaches and athletes about "strength" and they'll probably mention a rock-solid core. The six-pack abs are an emblem of splashy fitness. But the "core" remains a nebulous concept, with no scientifically agreed-upon definition. "There's so much mythology out there about the core," says Stuart McGill, Ph.D., a highly regarded professor of spine biomechanics at the University of Waterloo, in Ontario, Canada, and a back-pain clinician who's treated countless athletes with ruined lower backs.

Most researchers consider the core the corset of muscles and connective tissue that encircles and holds the spine in place. If your core is stable, your spine remains upright while your body pivots around it. Strong core muscles provide the necessary spinal stability, but only to a point. Think of the spine, Dr. McGill says, as a fishing rod supported by muscular wires. If all of the wires are tensed equally, the rod stays straight. "Now pull one of the wires really tight," he says. "What happens?" The rod buckles. So, too, he says, can your spine if you overwork the deep abdominal muscles.

Research at his lab and elsewhere has shown that repeated bending of the spine, such as that which occurs when most of us do crunches, can actually contribute to damage of the spinal discs. When cadaver pig spines were placed in machines and bent and flexed hundreds of times, the pigs' spinal disks almost always ruptured eventually.

So "go easy on crunches," Dr. McGill says, easy advice to follow for those of us who hate the things. "I treat an awful lot of patients," he adds with a sigh, "who have six-pack abs and a ruined back." (See the end of the chapter for advice on how to perform crunches correctly.)

Or forgo crunches altogether. Findings about the actual, measurable effects of core strength on athleticism have been mixed. A study several years ago of collegiate rowers, for instance, found that after eight weeks of an arduous core-exercise regimen—on top of their normal workouts—the rowers had great-looking abs but weren't better rowers; their performance didn't budge much at all in a rowing-machine time trial.

On the other hand, in a study that involved novice adult runners, each of whom displayed weak core strength to start with, those who completed six weeks of core training drills lowered their 5K run times more than a control group of beginning runners who did not focus on their midsections.

In essence, the strength we need depends on the strength we currently lack, and probably requires a concentration on "functional fitness," Dr. Nesser says. What each of us really needs to develop, he says, is enough strength "to make our bodies function better" in our sport and in life itself.

Forget the Formulas

There was a time when army basic training required endless sit-ups and frequent long runs in sleet while wearing military boots. The boots were the first element of that old-fashioned regimen to go. Now the long runs and the sit-ups are being phased out, too. The army is rethinking military strength and how to build it.

"We had to take a hard look" at how to sculpt soldiers, Lieutenant General Mark Hertling told reporters when describing the army's new approach to training. Lt. Gen. Hertling, who has a master's degree in physiology, spearheaded the effort, which soon will apply to all recruits. Out are sit-ups, deemed too injurious and ineffective. In are revised crunches, along with almost all yogic poses. Soldiers twist, plank, and bird-dog (in which they perch on all fours, extending an opposing arm and leg, like a pointer spotting quail). They complete wind sprints instead of ten-mile runs. The new regimen, Lt. Gen. Hertling said, is designed to better prepare soldiers for the actual demands of the field, as in this description of the humbling actions of Corporal Todd Corbin (from the Defense Department's "Heroes" website): "Running through the line of fire, he grabbed his wounded patrol leader and threw him over his shoulder. He then sprinted back to his Humvee, firing at enemies as he ran."

In a much less exalted manner, the rest of us can learn from the army's fitness evolution. The best strength training, science, and an increasing amount of real-world experience emphasize, is flexible, individualized, and not conducted in military boots. It does not need to be, as it once was, a rigid march from weight machine to weight machine, with unyielding prescriptions about how much weight to be lifting and when to progress. I know grown men who still shudder as they remember their high school football coach shouting at them to "raise the damn RM."

We can blame a famous and much-cited study from the 1960s for that, in which a group of inactive college students began following a variety of separate weight training regimens to see which produced the greatest strength gains. Before beginning any, the students determined their one-repetition maximum, or the most weight that they could lift or pull down with one grunting thrust or yank. The scientists used this number, familiarly known as the 1RM, to set the parameters of the subsequent routines. In some, the students lifted up to 90 percent of their 1RM for both upper- and lower-body exercises. In others, they used 70 or 50 percent of the 1RM. The researchers concluded that three sets of eight exercises performed at 85 percent of each individual's 1RM, with the total weight load being increased by 10 percent each week, led, after four weeks, to the greatest increases in strength.

The 1RM measure became standard among many weight trainers, but Dr. Phillips thinks that was a mistake. "Weight training does not have to be that complicated," he says. Recent studies at his lab have shown that people actually gain more strength by listening to . . . their bodies rather than following a formula. Does the lifting feel hard, especially after eight or ten repetitions? Then you're probably doing enough and should maintain that effort for a while. When lifting that same amount of weight feels easy, you need to do more. Don't worry about your 1RM.

There are a few other broad guidelines to maximize your benefits from strength training, says Dr. Phillips. In general, lifting less weight more times produces greater strength gains than the reverse. A study in his lab using college-age young men showed that they improved their maximal strength significantly more if they lifted less weight at least ten times than when they lifted more weight six to eight times. The additional repetitions seemed to stimulate greater changes within the muscle cells and the nervous system.

Not that you have to use formal weights or weight machines at all. Body-weight training, in which you rely on the formidable mass of your body—by which I mean no insult; almost all of us weigh something into the three figures, which is more poundage than most of us would lift at the gym—provides quite a workout. I haven't belonged to a gym in ten years. In part that's because I'm cheap. But it's also because I've learned that I carry my 120-plus pounds of resistance with me everywhere, even if I wish it were less.

Push-ups, pull-ups, squats, planks (a fine exercise before it became a silly Internet meme), and other exercises that use your body to provide resistance are do-anywhere, do-anytime moves that prompt the same cellular and aesthetic changes as using Nautilus machines. Try it. Drop to the floor in your office or living room right now and see how many push-ups you can complete. (No putting your knees on the floor.) Many physiologists and physicians consider a push-up a fine test of overall muscular fitness. According to national averages, a forty-year-old man should be able to complete about twenty-seven push-ups before becoming exhausted, and women that age should be able to do sixteen. I'm barely average on that measure.

Asanas and Squats

But I can downward dog with the best of them, and that's a means of gaining strength, too. Yoga and Pilates, as well as their many variants—Nia, Zumba, ballet aerobics, pole dancing, or what have you—have been shown to prompt muscular remodeling almost as readily as working with weight machines does. You apply substantial force to muscles when you stand on your head or balance your pretzeled legs and torso above the mat using only tensed arms. A group of college students at the University of California, Davis, who practiced an hour of yoga several times a week gained 30 percent more muscular strength after two months than they'd had when they started. Given the accompanying and agreeable toning of their muscles, their social lives probably improved, too.

Pilates, an elaborate series of body-weight exercises developed decades ago by the German bodybuilder and gymnast Joseph Pilates, also has scientific backing as a strength training alternative. In one study, eight weeks of Pilates training resulted in significant increases in leg and back strength among a group of middle-aged women who'd never practiced strength training before.

If you choose Pilates, yoga, or any method of strength training, however, and you're unfamiliar with proper form, seek out a qualified instructor. Ask about credentials, Dr. McGill says. And stop if any of the exercises or poses hurt. "I know plenty of yoga practitioners who've ruined their backs," he says. Be prudent. Start slow.

And don't look for effortless gains. Strength does have to be earned. Remember the fad, still continuing, of rocker-bottom "toning" shoes that promised painlessly to tighten muscles in your calves, thighs, and buttocks while you walked? ("Your boobs will be jealous," a refined advertising campaign for one of the shoe brands declared.) They didn't work. A study of muscle activation in women's legs while they wore the shoes proved that their muscles were contracting no differently than in ordinary shoes. No increased contractile force means no strength gains.

But those special shoes can benefit you, if you slip them on and then squat. Yes, squat. Not one of nature's more gainly movements, the squat is nevertheless extremely effective as a body strengthener. "I would nominate the squat as the single best exercise," Dr. Phillips says. "It's simple. It's convenient. It activates the body's biggest muscles, those in the buttocks, back, and legs." And it requires, in his version, no gym or coach—not even weights. "Just fold your arms across your chest, bend your knees, and lower your trunk until your thighs are about parallel with the floor," he says. "Do that twenty-five times. You don't need to do anything else. You probably won't be able to." Add a weighted barbell once the body-weight squats grow easy.

"It's a very potent exercise" for health, fitness, and physical performance, he concludes, and encapsulates everything you could wish for from strength training as a whole. "It builds power," he says. "It allows for progression. It's not overly complex. And you can do it anywhere."

Gaining Strength and Power Can Be Easy. Some Tips

1. First, Determine How Strong You Are.

You can go to a gym and measure your 1RM if you so desire. Or try the simple push-up test: Drop to the floor and perform as many full-body push-ups as you can. (Don't allow your knees to rest on the floor.) According to national averages, a 40-year-old man should be able to complete about 27 push-ups before becoming exhausted; women of the same age should be able to do 16. Add about 10 repetitions for every decade backward from 40, and subtract 5 for every decade forward. If you can't meet the averages, you should consider strength training, including regular push-ups. Couldn't do even one standard push-up? Then start gently by doing the exercise against a countertop at a 45-degree angle. As that becomes easy, move to stairs and then the floor.

2. Find an Expert.

If you've never done any strength training and want to begin a regular program, consider consulting a trainer with certification from the American College of Sports Medicine or the American Council on Exercise, and have him or her guide you through a workout, particularly if you plan to use free weights or machines.

3. Go Easy.

If you are your own strength coach, know that the latest science on the topic suggests that more repetitions with lighter weights will yield greater strength increases than heavier weights and fewer repetitions.

4. Crunch Carefully.

If you do crunches, do them correctly. Lie down, one knee bent and hands positioned palms down beneath your lower back for support, says Dr. McGill. Do not press your back against the floor or pull in your stomach. Imagine that your head and shoulders are lying on a bathroom scale. Lift them just a few inches, enough to move the imaginary scale to zero, hold briefly, and relax back down. Repeat eight to ten times, switch legs, and complete another eight to ten. Two other exercises that Dr. McGill recommends to complete a perfectly adequate core routine for most of us:

- **The side bridge:** Lie on your side, with your legs bent at the knee and your upper arm across your chest. Bend your lower arm so that your elbow is pointing away from your chest, with your other hand on your chest or hip. Slowly raise your shoulders, keeping your spine straight, and hold for eight to ten seconds. Repeat on the other side. After a few weeks, do the exercise with your legs straight. As even that exercise becomes easy, hold the position on each side for longer periods of time.

- **The bird dog:** Start on all fours, then slowly lift your right arm and left leg until each is parallel to the floor. Hold for eight seconds. Repeat with the opposite arm and leg. Do twenty reps (ten on each side). Keep your spine straight, hips level, and abdominal muscles slightly contracted. And don't forget to breathe.

5. Hop to It.

Plyometrics is another style of strength training, which involves leaping and hopping, creating explosive power in the process. As a side benefit, plyometrics likely improves skeletal health. In studies undertaken in Japan a few years ago, mice that jumped up onto a box (apparently rodents like plyometrics) forty times during a week increased their bone density significantly after twenty-four weeks, a gain they maintained by hopping up and down only about twenty or thirty times each week after that. Creating your own plyometrics routine can be relatively easy and involves little equipment or expertise. Some useful (and not too difficult) plyometric moves:

- **Vertical jumps.** Stand with your feet shoulder-width apart, your back straight, and your midsection flexed slightly forward. Lower your body until your thighs are parallel to the ground, and explode upward, jumping as high as possible. Land on both feet and explode again for the second jump. Stay on the ground for as short a time as possible.

- **Stair jumps.** Stand facing a staircase with your feet shoulder-width apart. Lower your body into a squat and jump onto the first stair. Land gently on both feet and as quickly as you can, jump up onto the next stair, until you reach the top (or get tired).

- **Tuck jumps.** Begin in a standing position with feet shoulder-width apart. Squat and then jump up, pulling both knees to your chest. Drop back, landing softly on both feet. Jump up again as quickly as you can.

7

When Bad Things Happen to Good Workouts

You can't be a runner past forty, as I am, and not have people constantly telling you that you are going to ruin your knees. They mean, presumably, that running will cause arthritis in the joint. It's not an unreasonable idea; other sports have been linked with early-onset arthritis in the knees. In a famous British study, almost half of a group of middle-aged, once-elite soccer players were found to have crippling bone-on-bone arthritis in at least one knee. Former weight lifters also have a high incidence of the condition, as do retired NFL players, although arthritis may be the most benign of their physical debilities.

But running is in fact probably not a problem for many knees. In a study that appeared not long ago in the scintillating European journal *Skeletal Radiology*, researchers with the Danube Hospital, in Austria, scanned the knees of a group of thirtysomething male marathon runners using MRI imaging both before and after the 1997 Vienna City Marathon. Ten years later, they scanned the same group of runners' knees. All but one of the men had continued to run marathons throughout the intervening decade, even as they entered middle age.

And, despite the hours of training, the scans showed that their knees had remained youthful. "No major new internal damage in the knee joints of marathon runners was found after a 10-year interval," the researchers reported. Only one of the participants had severely damaged knees, and he'd quit running before the 1997 marathon but was included in the study almost as an afterthought. His 1997 MRI showed a knee that was already in trouble, with cartilage lesions, swelling, and other abnormalities, some of which probably were congenital. In the decade that followed, his knee worsened substantially, accumulating additional tissue damage and cartilage lesions. His scan results prompted the researchers to speculate that he might have been better off if he'd kept running, since by the state of the other marathoners' still-sturdy, healthy joints, "continuous exercise is protective, rather than destructive to knees."

Why Me?

Why do some people become injured as a result of activity while others don't? That question should be of pressing interest to all of us who work out, because so many of us will be felled at some point. Depending on which statistics and studies you turn to—and how you define *injury*—between 30 and 90 percent of runners will hurt themselves in any given year. The incidence isn't necessarily much lower in other sports. Basketball players go down on the court all the time. Tennis players, soccer players, cyclists, swimmers, weight lifters, and yoga masters are also often sidelined.

The causes behind those painful statistics are myriad and, to a surprising degree, mysterious. Being inexperienced at a sport or activity predisposes you to injury, which makes sense. Newcomers are prone to mistakes in their training or technique that end in tweaked muscles or sore joints. Surveys of marathon runners have found that it is common for half of first-time racers to miss the starting line due to training injuries. But expertise isn't anodyne. Almost as many experienced marathon racers typically hurt themselves in the lead-up to the event. The overall percentage of elite runners injured in any given year is higher than the incidence among recreational runners.

Being of either of the two available genders doesn't provide much protection, either. Men numerically incur far more injuries than women in most activities, barring synchronized swimming and ballet. But far more men participate in most of the sports and recreational activities for which statistics exist. So the relative risks are different. Case by case, more men than women will tear an anterior cruciate ligament this year. But the percentages are much higher on the distaff side. In general, a woman is more likely to hurt herself while exercising than a man.

Does noting that distinction make this an opportune moment to discuss estrogen? There is more and more of it in sports, thankfully, but it remains little regarded scientifically. Meanwhile, it undeniably has an influence on injury risk. Some studies have found that postmenopausal women who take estrogen replacement have healthier muscles than those who do not. Even more striking, in a series of experiments for which not a single male with whom I am familiar would have volunteered, Canadian researchers gave estrogen to male athletes and then had them complete strenuous bicycling sessions. The men seemed to have developed entirely new metabolisms and muscles. They burned fuel differently during the ride than they had before and showed altered post-exercise markers of muscle development. There was no word, though, on any change in their response to repeats of *Sex and the City*.

The point, of course, is that being a woman affects your injury risk, particularly as estrogen ebbs and flows. An Australian study found that when women's estrogen levels were at their highest, around the time of ovulation, they landed subtly differently while hopping than at other times of the month. Their feet splayed, the arch collapsing just a little bit more than when their estrogen levels were lower. The women also seemed, to a small degree, wobblier. "We contend that the changes in foot biomechanics may be due to the effects of estrogen on soft tissue and/or the brain," says Adam Leigh Bryant, a senior lecturer at the University of Melbourne and author of the study.

Which does not suggest that female athletes are in some indefinable way more fragile than their male counterparts. Quite the reverse may be true, according to new research that I plan casually to mention as often as possible before my male training partners. In experiments at the University of Copenhagen in Denmark, scientists found that during exercise training, women's tendons and ligaments did not grow as thick and powerful as men's did, which had been expected. But after both men and women reduced or stopped their workouts, the women did not lose their training benefits as quickly. Estrogen, the researchers concluded, had maintained the women's hard-won strength and fitness gains better than men's bodies had held on to theirs, for a simple evolutionary reason. It was providing survival insurance, protecting the women "against fast muscle and collagen loss when she is inactive," as during pregnancy, the study's lead author, Mette Hansen, Ph.D., a researcher at the University of Copenhagen, told me.

The same evolutionary imperative seems to be at work in how women sweat.

Not long ago, researchers in Japan recruited a pool of trained athletes, male and female, as well as an age- and gender-matched group of untrained volunteers. All rode stationary bikes in a physiology lab heated to a balmy 86 degrees. The beginning of the hour-long session was leisurely; the pedaling intensity was only about 30 percent of each volunteer's VO2 max. Then it became tougher, rising to 50 percent of each rider's VO2 max, while the final 20 minutes required that the riders work at a strenuous 65 percent of VO2 max. Throughout, researchers monitored how much perspiration the cyclists were producing and how many of their sweat glands were active and pumping.

The fit men, unsurprisingly, perspired the most, but not because they were using more sweat glands. The fit women had as many active glands but produced less sweat from each gland. And the unfit women, by a wide margin, perspired the least, especially during the strenuous cycling, and became physiologically hotter before they began to sweat at full capacity. These results, the scientists concluded, "revealed a sex difference" in "the control of sweating rate to an increase in exercise intensity." In other words, the women were less adept at ridding themselves of body heat by drenching themselves in sweat.

Part of the reason seems, again, to be estrogen—and also testosterone, says Timothy Cable, Ph.D., a professor of exercise physiology at Liverpool John Moores University, in England, who has extensively studied exercise and sweating, an important issue in injury causation. In those same experiments in which brave men were injected with estrogen, they began to sweat less during exercise and instead mist delicately.

The gulf between how men and women sweat does not begin until puberty. Until then, the sweat rates of boys and girls are roughly the same. Then the sex hormones kick in. Even then, men's and women's perspiration is similar in terms of water, salt content, and smell, Dr. Cable says, strange as that may seem to anyone who's been in close proximity to a young male athlete's laundry hamper. (Sweat itself is odorless; the distinctive smell associated with sweating is produced by the waste products of bacteria feeding on the perspiration.) Men just produce more.

In practical terms, the sweat differential does mean that "women can be at a disadvantage when they need to sweat a lot during exercise in hot conditions," says Yoshimitsu Inoue, Ph.D., a professor at Osaka International University and one of the authors of the study. On the other hand, it may be that during evolution women had the good sense to get out of the hot sun, and their bodies adapted accordingly. The "lower sweat loss in women may be an adaptation strategy that attaches great importance" to preserving body fluids for survival, Dr. Inoue told me, while "the higher sweat rate of men may be an adaptation strategy for greater efficiency of action or labor."

Dr. Cable agrees. "Prehistoric men followed the herds," he says, whatever the temperature, while the women, cleverly, sought out the shade. "It's not a bad survival strategy," he says, even today.

Mad Dogs, Englishmen, and Muscle Cramps

But most of us, male or female, do not seek shade, and as a result heat illnesses contribute notably to the toll of disrupted workouts every year. "The body has a critical core temperature," Dr. Cable says, which occurs at about 104 degrees, after which the brain simply "shuts down the motor cortex." Unbidden, your legs stop churning and you curl up on the sidewalk until your core temperature drops (or a kind passerby calls 911). Sweating delays the onset of this critical heat buildup by dissipating the excess heat through evaporation.

Being in shape means that you begin to sweat at a lower body temperature, so being fit is a key to remaining fit while working out in the heat, says heat illness expert Dr. Douglas Casa. Acclimating slowly to soaring temperatures is also important, for both men and women, no matter how fit they may be.

So use your head. Cooling the neck before exercise in hot, humid conditions seems to be surprisingly effective in terms of competitive performance. Research conducted in steamy London found that healthy young men could run farther and faster in the heat if they first strapped on an ice-cold neck collar. The collar, lined with flexible artificial-ice packs, lowered the skin temperature on the men's necks and left them feeling less hot, even as their body temperature rose. The collars apparently cooled the blood in the neck's carotid artery, which then flowed to the brain to produce a "subsequent lowering of cerebral temperature," convincing the brain that the body was cooler than it really was and could push harder, says Christopher James Tyler, Ph.D., a lecturer in sport and exercise physiology at Roehampton University and author of the study.

Of course, as a general rule, tricking your body is not a wise injury-prevention strategy. But in this case, none of the athletes reached a dangerous core body temperature. They merely drew closer to that point than they otherwise might have, Dr. Tyler says.

Should you be planning to compete in hot weather, then you might want to create a DIY collar using a frozen bandanna (although the cold slap to the neck from cold collars has been known to cause an ice-cream headache, Dr. Tyler says). If you do use a neck cooler, monitor your physiological response carefully, Dr. Tyler advises. Cold necks did not change his volunteers' heart rates; they rose as would be expected. So if you find that "you're exercising at fifteen beats per minute more than normal, you might want to slow down."

You might also want to carry a few ounces of pickle juice. Exercise-related muscle cramps are common in hot weather, although probably not because the weather is hot. The fundamental cause of muscle cramps is in fact one of the continuing mysteries of physiology. Extremely pervasive, they can afflict anyone. We've all seen Olympians pull off the running track clutching a cramping calf muscle. I've been on bike rides with friends whose backs have begun to spasm. That was before I knew about pickle juice.

For years, most people, inside and outside academia, believed that cramping was caused by sweating-induced dehydration and the accompanying loss of sodium and potassium. Sufferers were advised to load up on potassium-rich bananas or chug large amounts of salty sports drinks. But that theory seems increasingly implausible.

Cramping athletes given fluids often continue cramping, while in an interesting experiment, college students who amiably agreed to have cramps induced in their big toe were unaffected by hydration. Whether the volunteers were well hydrated or dehydrated, scientists could induce the cramps, meaning that the spasms "were likely not caused by dehydration," says Kevin C. Miller, Ph.D., a professor in the Athletic Training Education Program at North Dakota State University, in Fargo, who oversaw the experiment.

Instead, he believes, muscle cramps probably are the result of muscular exhaustion and a cascade of accompanying biochemical processes. Certain mechanisms within muscles have been found to start misfiring when a muscle is extremely tired. Small nerves that should keep the muscle from over-contracting malfunction, and the muscle bunches when it should relax.

Which is where pickle juice enters as a palliative. In another of Dr. Miller's experiments, ten healthy male college students rode on specially configured semi-recumbent bicycles, set up so that only one leg pedaled. The laboratory was warm, so the men sweat. Each man cycled in thirty-minute bouts (with five minutes of rest between) until each lost 3 percent of his body weight through perspiration, a widely accepted definition of mild dehydration. The tibial nerve in the men's ankles was then electrically stimulated, causing a muscle in the big toe to cramp. (The procedure is too painful to use on larger muscles, like the hamstrings or the quadriceps, as anyone who has ever experienced a cramp in one of those muscles can understand.) The volunteers were told to relax and let the cramps run their course, which typically required about two and a half uncomfortable minutes.

Then their tibial nerve was zapped again. This time, though, as soon as the toe cramps began, each man downed about 2.5 ounces of either deionized water or pickle juice, strained from a jar of ordinary Vlasic dills. The reaction, for some, was rapid. Within about eighty-five seconds, the men drinking pickle juice stopped cramping. But the cramps continued unabated in the men drinking water. Pickle juice had "relieved a cramp 45 percent faster" than drinking no fluids and about 37 percent faster than water, Dr. Miller wrote.

Just how pickle juice performs its magic is still unknown. "The pickle juice did not have time" to leave the men's stomachs during the experiment, Dr. Miller points out. So the liquid itself could not have been replenishing lost fluids and salt in the affected muscles. Instead Dr. Miller believes that something in the acidic juice, perhaps even a specific molecule, may be lighting up specialized nervous-system receptors in the throat or stomach, which, in turn, send out nerve signals that somehow disrupt the melee in the exhausted muscle. It may be all down to vinegar. In one recent case report by other researchers, an athlete's cramping was relieved more quickly when he drank pure vinegar (without much pleasure, I'm sure) than when he drank pickle juice.

But for now, pickle juice is gaining adherents. In a survey of collegiate athletic trainers, a quarter of them said that they regularly dispense pickle juice to cramp-stricken athletes and that, in their experience, the stuff quickly brakes the cramping. So if your calf or other muscle suddenly, painfully catches, "try stretching it," Dr. Miller says. Doing so has been found in laboratory studies to significantly shorten the duration of a muscle cramp, most likely by shaking up and resetting the misfiring muscle and nerve reflexes. Or drain your emergency flask of Vlasic juice. It's not as palatable as bananas, Dr. Miller says, but unlike them, "it seems to work."

The Trouble with Shoes

The best thing that can be said of an exercise-induced cramp, though, is that it will soon dissipate, whatever your response. A cramp is a transient, acute ache. I'm not trying to minimize the discomfort; I've had a hamstring cramp. I used language that would scandalize a longshoreman. But the episode was over in minutes.

Overuse injuries are more lasting, debilitating, and demoralizing. They also are extraordinarily common, especially among runners. And for many of us, and I include myself in this group, our gear may be partly to blame. We're probably wearing the wrong shoes, and we can thank the U.S. Army for obligingly pointing this problem out to us.

A few years ago, the military began analyzing the shapes of recruits' feet. Injuries during basic training have been rampant for decades, but have been on the rise. Overweight, out-of-shape recruits don't help. But military authorities had hoped that by fitting soldiers with running shoes designed for their foot types, injury rates would drop.

Trainees obediently began clambering onto a high-tech light table with a mirror beneath it, designed to outline a subject's foot. Evaluators classified the recruits as having high, normal, or low arches, and they passed out running shoes accordingly.

You probably have had a similar experience. I know I have. For decades, coaches and shoe salesmen have visually assessed runners' foot types to recommend footwear. Many of us have stood on white paper, while a shoe salesman actually drew around our naked foot. Runners with high arches, like mine, were then directed toward soft, well-cushioned shoes, since it was thought that high arches prevented adequate pronation, or the inward motion of your foot and ankle as you run. Pronation dissipates some of the forces generated by each stride.

Flat-footed, low-arched runners (which I seem to be becoming as I age) tend to overpronate and have typically been told to try sturdy "motion control" shoes with firm midsoles and Teutonic support features, while runners with normal arches have been put into neutral shoes, which are often called "stability" shoes by the companies that make and categorize them.

But as the military was preparing to invest large sums in more arch-diagnosing light tables, someone thought to ask if the practice of assigning running shoes by foot shape actually worked. Military researchers checked the scientific literature and found that no studies had been completed that answered that question, so with admirable gumption they decided they would mount their own. They began fitting thousands of recruits in the army, air force, and Marine Corps with either the "right" shoes for their feet or stability shoes.

Over the course of three large studies, the researchers found almost no correlation at all between wearing the proper running shoes and avoiding injury. Injury rates were high among all the runners, but they were highest among the soldiers who had received shoes designed specifically for their foot types. If anything, wearing the "right" shoes for their particular foot shape had increased trainees' chances of being hurt.

Scientific rumblings about whether running shoes deliver on their promises have been growing lately. In another study, eighty-one experienced female runners were classified according to their foot type. About half of the runners received shoes designated by the shoe companies as appropriate for their particular foot stance (underpronators got cushiony shoes; overpronators, motion-control shoes; and so on). The rest received shoes at random. All of the women started a thirteen-week half-marathon training program. By the end, about a third had missed training days because of pain, with a majority of the hurt runners wearing shoes specifically designed for their foot postures. Across the board, motion-control shoes were the most injurious for the runners. Many overpronators, who in theory should have benefited from motion-control shoes, complained of pain and missed training days after wearing them, as did a number of the runners with normal feet and every single underpronating runner assigned to the motion-control shoes.

An influential article published a few years ago in the *British*

Journal of Sports Medicine concluded that sports-medicine specialists should stop recommending running shoes based on a person's foot posture. No scientific evidence supported the practice, the authors pointed out, adding that "the true effects" of today's running shoes "on the health and performance of distance runners remain unknown."

The lesson of the newest studies is obvious, if perhaps disconcerting. "You can't simply look at foot type as a basis for buying a running shoe," says Dr. Bruce H. Jones, the manager of the Injury Prevention Program for the United States Army Public Health Command and author of the military studies. The widespread belief that flat-footed, overpronating runners need motion-control shoes and that high-arched, underpronating runners will benefit from well-cushioned pairs is quite simply, he adds, "a myth."

The mythology grew and persists because "in certain aspects, the shoes do work," says Michael Ryan, Ph.D., the lead author of the study of female half-marathoners. Motion-control shoes do control motion, he says. Biomechanical studies of runners on treadmills repeatedly have proved that pronation is significantly reduced in runners who wear motion-control shoes.

The problem is that "no one knows whether pronation is really the underlying issue," Dr. Jones says. It's not clear how or even if over- or underpronation contributes to running injuries.

If you're heading out to buy new running shoes, then, be your own best advocate. "If a salesperson says you need robust motion-control shoes, ask to try on a few pairs of neutral or stability shoes, too," Dr. Ryan says. "Go outside and run around the block" in each pair. "If you feel any pain or discomfort, that's your first veto." Hand back those shoes. Try several more pairs. "There really are only a few pairs that will fit and feel right" for any individual runner, he says. "My best advice is, turn on your sensors and listen to your body, not to what the salespeople might tell you."

Baring It All

Or go naked.

At a recent symposium hosted by the American College of Sports Medicine and cutely titled "Barefoot Running: So Easy, a Caveman Did It!" a standing-room-only crowd waited expectantly as a slide flashed up posing this question: Does barefoot running increase or decrease skeletal injury risk?

"The answer," says Indiana University's Dr. Stuart Warden, "is that it probably does both."

Barefoot running is about as trendy at the moment as any millennia-old activity can be. Books and websites evangelize about barefoot running, with proponents promising that it's a more "natural" way to run and will vastly reduce the chances of injury. "There are people who are convinced that barefoot runners never get injured," says Daniel E. Lieberman, a professor of human evolutionary biology at Harvard, who runs barefoot himself and spoke on the topic during the symposium. "That's not the case."

Instead, evidence has mounted recently that some runners, after kicking off their shoes, have wound up hobbled by newly acquired injuries. These maladies, instead of being prevented by barefoot running, seem to have been induced by it.

What happens to a modern runner when he or she trains without shoes or in the lightweight, amusingly named "barefoot running shoes" is in fact an object lesson in injury mechanics. Most of us, after all, grew up wearing shoes. Shoes alter how we move. An interesting review article in the *Journal of Foot and Ankle Research* found that if you put young children in shoes, their steps become longer than when they are barefoot, and they land with more force on their heels. Similarly, when Dr. Lieberman traveled recently to Kenya for a study he later published in the journal *Nature*, he found that Kenyan schoolchildren who lived in the city and habitually wore shoes ran differently than those who lived in the country and were almost always barefoot. Asked to run over a force platform that measured how their feet struck the ground, a majority of the urban youngsters landed on their heels and generated significant ground reaction forces or, in layman's terms, pounding. The barefoot runners typically landed closer to the front of their feet and lightly, without generating as much apparent force.

Based on such findings, it would seem as if running barefoot should be better for the body, because less pounding should mean less wear and tear. But the problem is that, for better or worse, the body stubbornly clings to what it knows. Just taking off your shoes does not mean you'll immediately attain proper barefoot running form, Dr. Lieberman says. Many newbie barefoot runners continue to stride as if they were in shoes, landing heavily on their heels.

The result can be an uptick in the forces moving through the leg, Dr. Warden pointed out, since you're creating as much force with each stride as before but no longer have the cushioning of the shoe to help dissipate it. Most barefoot runners eventually adjust their stride, landing closer to the front of their feet—since landing hard on a bare heel hurts—but in the interim, he says, "barefoot running might increase injury risk."

Even when a barefoot runner has developed what would seem to be improved form, the forces generated remain potentially injurious. In a study from the biomechanics laboratory at the University of Massachusetts, runners strode across a force plate, deliberately landing either on the forefoot or on the heel. When heel striking, the volunteers generated the expected thudding ground reaction forces; when they landed near the front of the foot, the force was still there, though for the most part it had a lower frequency, or hertz. Earlier research has shown that high-frequency forces tend to move up the body through a person's bones. Lower-frequency forces typically move through muscles and soft tissue. So shifting to a forefoot running style, as people do when running barefoot, may lessen your risk for a stress fracture but increase your chances of developing a muscle strain or tendinitis.

In other words, the "evidence is not concrete for or against barefoot or shod running," says Allison Gruber, who conducted the hertz study. "If one is not experiencing any injuries, it is probably best to not change what you're doing."

On the other hand, if you do have a history of running-related injuries or simply want to see what it feels like to run as most humans have over the millennia, then "start slowly," says Dr. Lieberman. Remove your shoes for the last mile of your usual run and ease into barefoot running over a period of weeks, he suggests, and take care to scan the pavement or wear barefoot running shoes or inexpensive moccasins to prevent lacerations. "Don't overstride," he says. Your stride should be shorter when you are running barefoot than when you are in shoes. "Don't lean forward. Land lightly." Humans may have been built to run barefoot, he says, "but we did not evolve to run barefoot with bad form."

Pretty Is as Pretty Does

The question of form, in almost any sport, is freighted. What is good form? Who decides? Will pretty form make you fast? Will it keep you free from injury? Anyone who ever has been a spectator at a major marathon knows that the world's best distance racers come in a variety pack of sizes and styles, with an accompanying wide range of running forms. Some racers are tall and lanky, with flowing, lissome strides. Others are more compact, their legs churning beneath them like pistons. And some are like British champion Paula Radcliffe, who runs like a praying mantis, all knees and angles, her arms crooked at odd angles and her neck jerking from side to side. All of these runners, no matter how ungainly or gorgeous their strides, are incalculably faster than the rest of us.

The same situation applies in most sports. Some of the world's most decorated swimmers, cyclists, tennis players, basketball players, and others have had idiosyncratic form. The Canadian Steve Nash used to bag free throw after free throw while tossing the ball underhanded. Australian swimming legend Ian Thorpe always kicked much deeper than any coach would recommend (a "mistake" mitigated by his size-200 feet), and Lance Armstrong famously sat high on the bike and rocked his hips during time trials, technical errors if employed by any other rider.

Given this wide disparity of styles and mechanics among successful athletes, is form something that the rest of us need to worry about? Does it matter, really, how we look? Many coaches and athletes believe it does. Paula Radcliffe may be fast, but she's also fragile and frequently injured, although to what degree her oddball form contributes to her injury history is unknowable.

Terrence Mahon, a coach who's worked with some of America's best distance runners, has his athletes practice form drills for hours at a time. "Proper form will lessen the chance of injuries and further the longevity of an athlete's career," he says.

But scientific studies are more equivocal. In one, researchers set out to examine "the consequences of a global alteration in running [form]" in triathletes. They wanted to see if dramatically altering the athletes' strides would make them more efficient, better runners. The athletes spent twelve weeks relearning how to run, with instruction on how best to hold their bodies and strike the ground with their feet. By the end, all of the triathletes involved had changed their running form, to a greater or lesser degree; most had become prettier runners, less bouncy, with slightly shorter strides. They also, for the most part, had become less efficient. It was now costing them more oxygen to cover a given distance than before they'd changed their form. Running, essentially, had become physiologically harder.

Similarly, researchers at several different labs recently determined, convincingly, that the most physiologically efficient cycling

form would involve a pedal cadence of about sixty strokes per minute. At that cadence, the riders get the most metabolic bang for the
buck. Pedaling faster or slower demands relatively more oxygen and
energy. But good luck finding a professional cyclist who turns his
pedals that slowly. Most maintain a cadence of around ninety pedal
strokes per minute. When one group of researchers had experienced
cyclists reduce their cadence to near sixty strokes per minute, most
of the riders showed measurable gains in physiological efficiency;
they were using less oxygen at a given cycling pace. But none enjoyed the experience and all returned to their former pedaling form
as soon as the study ended.

All of these studies underscore the difficulties inherent in defin-
ing a generally ideal form, in almost any sport. Even when science
can show that one method of performing a skill works better than
another, that style may not be achievable or comfortable for everyone. Frequently, even if a change in form offers benefits, it has side
effects, such as sore muscles and hobbling injuries. I'd direct you to
barefoot running again. When researchers in South Africa recently
taught a group of recreational runners to land near the middle of
their foot, barefoot-style, instead of on their heel, the subjects were
enthusiastic. In the early days of the experiment, most of the volunteers said they felt lighter and faster. They liked this new running
form. But within weeks, the same runners were reporting painfully
stiff and sore calf muscles and a spate of Achilles tendon injuries.
Not one got through the program without some kind of injury.

So is worrying about form ill-advised? Probably not. After all,
as Heisenberg might have told us, the very act of thinking about
how you move can have positive impacts, even if the actual changes
do not. In an intriguing study from Germany, researchers asked
experienced runners to concentrate on their form during treadmill
sessions. The runners were told to think about their arm and leg
movements and their breathing. Similar studies in which runners
were told to think about their thinking—concentrating on how

they felt or how far they still had to run, etc.—usually ended with the runners performing badly. But in this experiment, the runners improved their running economy and oxygen consumption compared with their baselines measures. When they thought about their form, running became physiologically easier.

Ultimately, form may be the wrong idea here. *Technique* or *mechanics* would be better terms. If form concerns how your body moves through space, as a whole, then technique and mechanics are about smaller, more discrete elements of movement, such as how you hold your head or position your backside on a bicycle or angle your ankles as you swim. There is more and better science about mechanical corrections than about global changes to form. The science suggests that, in many cases, small alterations to technique can make noticeable differences in how easy an activity feels or how adept you become.

In fact, expending excessive effort on tweaking your technique may be misdirecting your energies. When researchers examined runners who'd decided to take on longer races—moving up from the 10K distance, for instance, to the marathon or from the marathon to ultra-distances—they found that the best predictor of success in the longer races was not biomechanics, not how the runners ran, but how much they trained. A runner with an oddball stride who put in the miles was more likely to finish faster than a gorgeous strider who did not. Plan accordingly.

And stay safe out there.

The Injury Catch-22

"The single biggest predictor of injury is previous injury," says exercise physiologist Dr. Ross Tucker, who also founded a widely read exercise science blog. If you've never been hurt during your favorite activity, you're much less likely to be hurt than someone who has been hurt before. It rather seems as if perhaps nature should spread the pain around more. But injuries demonstrably cluster, and they will cluster in you if you've suffered an injury before.

To understand why, it helps to know what happens to bovine knee joints if you set about dropping weights on them from above. Knees, no matter how elastic your aesthetics, aren't beautiful. They look ungainly and cobbled together. And they're not much prettier on the inside, where a mad latticework of ligaments and tendons laces together multiple muscles and bones. Behind your kneecap, the pinky-sized anterior cruciate ligament diagonally joins and stabilizes the tibia and femur, the posterior cruciate ligament backs it up from behind, the lateral and medial collateral ligaments buttress the knee's sides, and the hamstring, quadriceps, and patellar tendons bring the leg's big muscles into play. No joint in the body is more important for smooth, powerful, and elegant movement. The knee, with its bony, sinewy, interwoven parts, absorbs and diffuses forces equivalent to at least ten times your body weight with every running stride. It gyrates and coils when you pivot, allowing your foot to point one way and your trunk another. Basketball would be inconceivable without the human knee; so would Baryshnikov.

It's this very complexity and ubiquity that also make the knee vulnerable. There are so many ways you can hurt it—a hard twist, a slight bump, a maladjusted bike seat, a "garage sale" on the ski slopes, or a misguided, no-preparation return to the soccer pitch at your twenty-year high school reunion. But what occurs at the cellular level inside a wrenched, wounded knee is ugly—and indicative of why one injury predisposes you to another.

In a rather medieval version of advanced science, Constance Chu, M.D., a professor of orthopedic surgery at the University of Pittsburgh and director of the Cartilage Restoration Center there, and her collaborators dropped a heavy weight onto parts of a cow's knee joint from various heights. (The cow was not present; the knees came from a local abattoir.) When the weight hit the joint's surface cartilage head-on with great speed and force, the bone and cartilage fractured and splintered. No surprise there. But that type of damage is rare in sports, unless a racecar hops a retaining wall and plows into spectators' knees.

More often the impacts that cause knee injuries in sports are subtle and glancing. So Dr. Chu continued her experiment, using lighter weights and lower overall force against the cows' knees—a better approximation of the perturbations inside a human knee when a ligament is torn.

At first, thumped more gently, the various parts of the knee appeared to be fine. But when Dr. Chu and her colleagues examined the cartilage cells just below the placid surface, they found carnage. "Many of the cells within the impact zone"—the area that had been lightly but directly thwacked by the weight—"were dead," she says. They had died instantly. More insidiously, other cartilage cells, those outside the injury site, began to die in the hours and days after the impact. "We saw an expanding zone of death," Dr. Chu says. By the end of her group's planned observation period, four days after the impact, cartilage cells well away from the original injury site were still dying.

In other words, what had begun as the equivalent of a knee-banging fall on the ski slopes had escalated at a deep, cellular level into widespread and ongoing damage. This is particularly worrying in knees, since, as Dr. Chu says, cartilage cells do not readily regenerate. The affected cartilage will be weakened permanently, making it more prone to tears and, as the knee components rub against one another during everyday movement, the development of arthritis. All from a relatively minor knee "insult."

Something similar, although more cerebral, occurs when you twist an ankle, which is, in fact, the single most common injury in sports. Back in the mid-1960s, a physician wondering why his patients with a previous ankle sprain often suffered another asked the affected patients to stand on their injured leg (after it was no longer sore). Almost invariably, they wobbled badly, flailing their arms and having to put their other foot down much sooner than people who'd never sprained an ankle. With this simple experiment, the doctor made a critical, if in retrospect seemingly self-evident discovery: People with bad ankles have bad balance.

"There are neural receptors in ligaments," says Jay Hertel, Ph.D., an associate professor of kinesiology at the University of Virginia and an expert on the ankle. When you damage the ligament, "you damage the neuro-receptors as well. Your brain no longer receives reliable signals" from the ankle about how your ankle and foot are positioned in relation to the ground. Forever afterward, your proprioception—your sense of your body's position in space—is impaired. You're less stable and more prone to falling over and reinjuring yourself.

No First Strike

So what's to be done? The science is unequivocal on this point. Don't. Get. Hurt. In the first place.

You can move on to the next chapter now.

Unless, of course, you have experienced a sports injury in the past or suspect that you're innately clumsy or wonder if there are any methods at all to inoculate yourself against a likely injury or somewhat reduce your risk of suffering that initial strain, sprain, tear, or relentless soreness.

And there are. "Besides a history of injury," the "other main risk" for injury involves "training errors," Dr. Tucker says. Rapidly increasing the amount or intensity of your training, switching surfaces, trying out new shoes—any one of those can predispose you to injury. So make adjustments slowly, listening for telltale creaks and complaints in your muscles and joints.

Practice balance training, too, no matter what your sport. Runners, both young and adult, have been shown to suffer more knee injuries from tripping over small obstacles than from any other cause. We can, I'm sorry to say, be a graceless bunch. "We have lots of equipment here in our lab" for patients to test, stress, and improve their balance, Dr. Hertel says. "But all you really need is some space, a table or wall nearby to steady yourself if needed, and a pillow." Begin by testing the limits of your equilibrium. If you can stand sturdily on one leg for one minute, cross your arms over your chest. If even that's undemanding, close your eyes. Hop. Or attempt all of these exercises on the pillow, so that the surface beneath you is unstable. "One of the take-home exercises we give people is to stand on one leg while brushing your teeth, and to close your eyes if it's too easy," Dr. Hertel says. "It may sound ridiculous, but if you do that for two or three minutes a day, you're working your balance really well." And amusing your spouse as a bonus.

The Riddle of the Tendon

If, despite all the scolding and nagging from science, you have managed to damage some body part due to exercise, be wary of easy succor. Many athletic injuries, especially overuse injuries of the tendons and ligaments, can be lingering and difficult to treat. They require time and rest to heal. Nothing else has been proven effective, and trying to rush nature can be counterproductive, as the cautionary tale of cortisone makes clear.

In the late 1940s, the steroid cortisone, an anti-inflammatory drug, was synthesized and hailed as a landmark. It provided a safe, reliable means to lessen the pain and inflammation associated with many injuries and became standard treatment for tennis elbow and aching Achilles tendons, among other conditions.

Then came the earliest clinical trials, including one, published in 1954, that raised niggling doubts about cortisone. In that early experiment, more than half of the patients who received a cortisone shot for tennis elbow or other tendon pain suffered a relapse of the injury within six months. Yet such experiments didn't slow the drug's ascent. It had such a magical, immediate effect against pain.

But numbing pain is not synonymous with healing tissues. A major review article of the effects of cortisone scoured the results of nearly four dozen randomized trials, which had enrolled thousands of people with tendon injuries, particularly tennis elbow and Achilles tendon pain. The reviewers determined that cortisone injections did provide fast and significant pain relief, compared with doing nothing or following a regimen of physical therapy. The pain relief could last for weeks.

But when the patients were reexamined at six and twelve months, those who'd received cortisone shots had a much lower rate of full recovery than those who had done nothing or undergone physical therapy. They also had a 63 percent higher risk of relapse than people who had adopted the time-honored wait-and-see approach. As Bill Vicenzino, Ph.D., the chairman of sports physiotherapy at the University of Queensland, in Australia, and author of the review, told me, of people with tennis elbow or other overuse injuries, 70 to 90 percent who follow a "wait-and-see policy" get better after six months to a year. But those getting cortisone shots "tend to lag behind significantly at those time frames." In other words, in some way, the cortisone impedes full recovery, and those getting the shots "are worse off," he says.

Those people receiving multiple injections may be at particularly high risk for continuing damage. In one study that the researchers reviewed, "an average of four injections resulted in a 57 percent worse outcome when compared to one injection," Dr. Vicenzino says.

Why cortisone shots should slow healing is a good question. An even better one, though, is why they help in the first place. For many years it was widely believed that tendon-overuse injuries were caused by inflammation, says Karim Khan, Ph.D., a professor at the School of Kinesiology at the University of British Columbia and the coauthor of a commentary accompanying Dr. Vicenzino's review. The injuries were, as a group, given the name "tendinitis," since the suffix -itis means "inflammation." Cortisone is an anti-inflammatory medication. Using it against an inflammation injury was logical.

But in recent years, numerous studies have shown, persuasively, that overuse injuries do not involve inflammation. When animal or human tissues from damaged Achilles or other tendons are examined, they don't contain biochemical markers of inflammation. Instead, the injury seems to be degenerative: The fibers within the tendons fray. That's why today the injuries usually are referred to as tendinopathies, or diseased tendons.

In the short term, cortisone, an anti-inflammatory, seems able to combat the pain of noninflammatory injuries by influencing the neural receptors involved in creating the soreness, Dr. Khan says. "The shots change the pain biology" within the tissue. But cortisone shots do not "heal the structural damage" underlying the pain, he says. Instead, they actually "impede the structural healing."

So whether cortisone shots make sense in treating abused, overused tendons depends on how you choose to "balance short-term pain relief versus the likelihood" of longer-term negative outcomes, Dr. Khan says. Some people, including physicians, may decide that the scale still tips toward cortisone. There will always be a longing for a quick, effortless fix, especially when the other widely accepted alternatives for treating sore tendons are to do nothing or, more onerous to some people, rigorously exercise the sore joint during physical therapy.

Still, think hard about your options if you have been hobbled with a sore Achilles tendon, elbow, or similar injury. Some emerging science suggests that moving and loading the sore part, if you are gentle and consistent, will spark molecular changes within the frayed tissue that eventually should end in healing. Within six months to a year, you could have a like-new tendon. That kind of slow, patient effort won't provide as immediate or showy a response as injections, but according to Dr. Vicenzino, it's likely to be more physiologically productive.

If It Ain't Broke

Finally, and perhaps most important, all of the available science suggests that if you've gotten this far with your current exercise regimen and haven't experienced many, if any, injuries, you should be safe or certainly at greatly reduced risk for problems in the immediate future. The body adjusts to your routine. That is the whole point of a physiological training response. Muscles become as thick and sturdy as required to carry you through your current workout. Ligaments and tendons become elastic enough but not too loose. Even seeming abnormalities or niggling aches may actually signal adjustments within your musculoskeletal system that will make you more resistant to injury, if you're careful. When researchers at Monash University, in Melbourne, Australia, compiled and reviewed several decades' worth of studies about activity and knee health, covering distance running, basketball, soccer, and several other sports, they found that, at first blush, strenuous physical activity did seem to damage knees. Activity, especially lots of it, was "associated with an increase in radiographic osteophytes," or bone spurs, the authors wrote, a condition that long has been accepted as an early indication of knee arthritis. Some of the studies under review had, in fact, concluded that activity must eventually end in arthritis, since the examined knees appeared to be imperiled.

But as the Australian researchers pointed out, some of those same studies, as well as others, did not find other characteristic changes in the knee that indicate damage. There was, for instance, almost no joint-space narrowing in active people. Joint-space narrowing is a necessary if unwelcome step on the way to full, bone-on-bone knee arthritis. The shock-absorbing cartilage in the joint wears away, the bones move closer together, and the space between tapers. Active people did not display this narrowing. In fact, according to a number of the studies reviewed, active people had greater cartilage volume than sedentary people. They weren't losing the tissue; they were vigorously maintaining it.

Why, then, were their knees so often sprouting bone spurs, supposedly a marker of damage? The answer may be that in an active person's otherwise uninjured knees, spurs are healthy, says Flavia Cicuttini, Ph.D., a professor at the School of Public Health and Preventive Medicine, at Monash. The spurs, she says, "may simply be a way that the bones adapt to forces pulling on the joint."

As confirmation, a large, ongoing examination at Stanford University has examined what factors allow people to live healthy lives well into their eighth and ninth decades of life. As part of that work, researchers began following a group of middle-aged recreational distance runners. They checked in with the runners periodically for nearly two decades, beginning in 1984, when most of the runners were in their midfifties or -sixties. At the start of the study, 6.7 percent of the runners had creaky knees, with mild symptoms of arthritis. None of an age-matched group of non-runners had any symptoms of arthritis. But after twenty years, 32 percent of the non-runners had arthritic knees, according to scans of their joints. Only 20 percent of the runners did, and barely 2 percent of their knees were severely arthritic. Ten percent of the non-runners' knees were.

"We were quite surprised," says Eliza Chakravarty, Ph.D., a Stanford professor who led the study. "Our hypothesis going in had been that runners, because of the repetitive pounding, would develop more frequent and more severe arthritis." Instead the elderly runners had healthier knees than the older people who didn't hit the roads. "But what most struck me," Dr. Chakravarty says, "is that the runners we studied were still running, well into their seventies and eighties. They weren't running far," she says. "They weren't running frequently. They averaged perhaps ninety minutes a week. But they were still running."

Strategies to Keep Yourself Off of the Injured Reserve List, Whatever Your Sport

1. Strengthen Your Knees

The best way to ensure that running doesn't hurt your knees or other body parts is to not hurt them in the first place. In study after study, the primary predictor of an athletic injury, to the knee or any other joint or tissue, is a previous injury. A proven deterrent against a first (or subsequent) knee injury is some targeted strength training. If you belong to a gym and know what you're doing, concentrate on exercises that strengthen the quadriceps and the hip stabilizers. For the rest of us, the following exercises are a scaled-down knee-protection routine that's easy to do in a living room or the nearest stairwell:

> • **Front Step-Ups.** Place one foot on the first step of a staircase or low box and stand straight up, keeping your weight on your supporting leg. Lower yourself back down, but don't shift your weight; just touch your heel to the floor. To make the exercise harder, raise your knee toward your chest while lifting your arms to a position parallel to the floor. Repeat with the other leg. Try to complete five repetitions on each leg to start, increasing the number of repetitions as the exercise becomes easy.

- **Wall Squats.** Stand with your back against a wall, knees slightly flexed, feet in front of you. Slowly slide your back down the wall until your knees are bent as close to ninety degrees as possible. Hold for 30 seconds, increasing the time to a minute as the exercise becomes easy. Straighten your knees and ease yourself back up the wall. Repeat the full exercise five times.

- **Straight-Leg Lifts.** Sit on the floor, back straight, one leg extended and the other bent toward your chest. In this position, lift the straight leg slightly off the ground. Lower it to the floor, then repeat five times. Switch to the other leg and complete five consecutive repetitions. Increase the number of total repetitions as the exercise becomes easy.

2. Better Form?

No one knows whether "ideal" form lessens injury risk; science hasn't found a link between pretty form and fewer injuries. But many coaches and athletes are convinced that the two elements are intertwined, and that performance may be improved as well.

So what follows are some broad tips on technique for runners, swimmers, and cyclists, based on science where available, and on the experience of practitioners and coaches where not. None of the suggestions will be right for every person, given the enormous disparity in our body types and abilities. Any should be implemented with caution. If a change in mechanics seems to be resulting in excessive muscle pain, stop doing it. There's no benefit to being benched.

For Runners

• Aim for "a straight, perpendicular line between the ground and your ankle and knee," elite running coach Terrence Mahon says. Most biomechanists and experienced coaches agree that how your feet strike the ground is probably less important than where. Your feet should be in line with your hips, not far out ahead of them. A good drill for improving foot positioning is known as "high knees." While striding forward, pull up one knee after the other until your thigh is parallel to the ground. Repeat as many times as possible in about fifteen seconds. Over time, try to increase the number of knee raises you can complete in that same fifteen seconds. Keep your back straight and make sure that your feet land square with your hips.

• If you don't have perfectly upright posture while running, lean slightly forward rather than back, especially as you push off from the ground. A 2010 German study of college-age runners found that leaning forward at the moment of toe-off resulted in faster running. Other, older studies have found that a slight forward lean (no more than about 10 percent) seems somehow to ease the effort of drawing in breath, which, in turn, can make running feel less strenuous.

• When striding, "imagine you're running on eggshells," Mahon says. Many of us thud with each step, landing with the grace and lightness of Lurch. Try, instead, at least periodically, to alight gently enough to avoid mashing imaginary eggs underfoot. The process can be salutary, highlighting the best ways for you, personally, to adjust your stride for milder collisions with the ground. It also can be diverting, helpful for passing those long, tedious middle miles in a training run.

• Don't overstride. Multiple studies of experienced runners have found that people tend naturally to develop a stride
length that is best for them physiologically. The body is clever
that way. But for unknown reasons, some people's strides are
slightly too long, which can send impact forces shooting
through the shinbone, resulting occasionally in stress fractures.
In computer modeling work at Iowa State University, in Ames,
for a study published in 2009, researchers attached reflective
markers to the bodies of ten collegiate cross-country runners
and had them dash repeatedly down a twenty-five-meter run-
way equipped with a force plate, while elongating or shortening
their natural stride. Computer programs calculated just how
much force was being applied to the shinbone under different
striding conditions and determined that reducing stride length
by about 10 percent would reduce the stress on the tibia enough
to substantially lower stress-fracture risk. Those with a history
of lower-leg stress fractures (one of the most common injuries
in sports) might try tweaking the length of their steps. Uncertain what a 10 percent change in your stride length would be?
"It's about as much as you can shorten your stride without it
beginning to feel quite uncomfortable," one of the study's authors told me. Absolute precision isn't necessary. "Seven or eight
or nine percent is fine," he says.

• Less shoe is probably best. In a large-scale study of army
recruits, the new soldiers were assigned shoes based on whether
they appeared to overpronate. Pronation, as you'll recall, is the
inward roll of your foot as it lands. Some people pronate more
than others, which was thought to contribute to injury risk, and
motion-control shoes, which are heavy and loaded with features
designed to reduce or prevent pronation, promise to lessen that
risk. But in the army study, the recruits who wore motion-
control shoes during basic training were much more likely to

wind up injured than other runners. The army scientists found the fewest injuries among those who wore a so-called neutral, or unflashy, lightweight shoe. Cheap is fine. In a recession-friendly study from Britain, low-priced running shoes provided the same cushioning as the highest-priced.

For Swimmers

Unlike running, swimming is not natural to humans, despite our long-ago aquatic roots; as a result, technique is especially important. You can run ugly and reach the finish line. Swim poorly and you may sink. (Presumably you know how to swim, or you wouldn't be reading this. If not, skip this section and, please, enroll as soon as possible in a local learn-to-swim program.) The following suggestions are aimed at recreational and competitive swimmers, of any stroke persuasion.

- Take it slow. "There's a lot of doing the same thing in swimming," Olympian Ryan Lochte says. "It helps to shake things up sometimes, to try different things." He does this by focusing on minute issues of technique during individual workout sessions. And he does that by swimming like the rest of us: slowly. "The only way to really work on technique is to swim very slowly and really think about every little thing that you're doing," he says. "How your body is positioned, what your hips are doing, the positioning of your shoulders and hands and feet." Ideally, he adds, you should "stay high in the water, don't fight the water, move with the water"—koans for all aquatic types.

  —To that end, work on keeping your belly above the surface of the water during the backstroke, and try swimming during some practices with a piece of buoyant foam (often called a pull buoy) between your legs. By using a buoy, you start to feel how your body ideally should be positioned in the water. Next, try to replicate that without the buoy.

• Streamline. You move faster under water than above it. "If a swimmer pushes hard off the wall and remains streamlined under water," says Gregg Troy, an elite swimming coach, "that means you'll transition into your stroke with much more momentum. It's almost as if you're swimming downhill. That's very important."

• Kick, baby, kick. Kicking workouts were once an afterthought for many swimmers. "It takes more time in practices to kick than to just swim," Troy says, so you get "less overall swimming volume," which in earlier days prompted many coaches to ignore the skill (to the relief, no doubt, of their charges; kicking is hard work). But today most coaches and swimmers have "come to realize that less volume with more kicking produces world records," Troy says.

—Try kicking 100 meters at a speed as fast as you can maintain, rest for 20 seconds, then kick for another 100 meters. Keep your pace consistent. Work your way up to 10 repetitions. Then do the same drill, only in 25-meter sprints. Finally, on a separate day, kick nonstop for 20 minutes, alternating hard 25-meter sprints with slower 50-meter recovery swims. Really concentrate on kicking technique. Experiment with slight tweaks to your ankle position.

—If you've never used fins, try a pair. A number of studies of the biomechanics of swimming have shown that regular use of fins during kicking drills can increase ankle mobility, which, in turn, should increase the power of your kicks when you remove the fins. Plus, if you've never used them, fins make you really fast, which, even if the sensation is short-lived, is glorious.

For Cyclists

One of the keys to avoiding injury in cycling isn't physiological; it's mechanical. Your bike must fit. "If a bicycle isn't fitted properly to a rider, that rider is going to have knee and back trouble," says Jonathan Vaughters, a former professional bike racer and team manager for top U.S. professional cycling teams. The following tips provide a basic outline of proper bike fit. If you're a racer or if you develop knee or back pain after riding, schedule a computerized bike-fitting session at a professional shop, which will be more precise and customized.

- For proper seat height, put your heel on the pedal and turn it until the crank arm is at the bottom of the pedal stroke. The seat is at the proper height when your knee barely locks; it will be slightly bent when your cleat is in place for a ride. More specifically, aim for a thirty-five-degree angle in your knee when your foot is at the bottom of the pedal stroke. When researchers recently compared several different seat heights for cycling economy, they found that the thirty-five-degree-angle setting was ideal for both power production and minimizing knee problems.

- For proper seat setback, imagine a straight line from the tip of your saddle to the bottom bracket. Edge your seat back from that line, so that the nose is about 0.05 inches behind the bottom bracket for every inch you are tall. Your back should be straight and your shoulders shouldn't feel hunched.

• For proper stem length, sit on the correctly positioned seat and look at the front wheel hub. The handlebars should obscure your view of the hub. If you can see it, you probably need a longer or shorter stem. Your bike shop should be able to swap stems for you. But if you wind up needing a stem shorter than nine centimeters or longer than thirteen centimeters, you may need to look for a different size bike.

• Your cleat, when locked into the pedals, should put your feet in a position that closely approximates how you walk (for instance, pigeon-toed). The ball of your foot should be directly above the pedal axle. For those who don't use clip-in pedals, consider switching to them. Science and experience both show that clip-in pedals provide efficiency and, surprisingly, safety. "You have far more control when you're clipped into your pedals than when you're not," Vaughters says. With practice, you will be able to unclip and put down your foot to balance yourself when stopping as quickly as with flat pedals. The learning curve is, admittedly, steep and humiliating: Practice on grass at first, to absorb the inevitable falls as you learn how to twist your foot and unclip. But the bumps will make you a better rider.

8

How to Build a Better Brain

The sea squirt is not one of nature's more charismatic creatures, but its life story is instructive to modern humans. Tubular, opaque, and squelchy, it resembles a worm-fish from Mars. But the sea squirt is in reality more closely related to humans than to other fish. It's a member of the chordate family, just as we were long ago, in another evolutionary form. When scientists sequenced the entire genome of the sea squirt a few years ago, they found long sections of DNA identical to our own.

At birth, infant sea squirt larvae have a brain. Not much of one; it consists of a few hundred brain cells and some nerve endings. But it does allow the squirt to think in a rudimentary fashion. Young squirts need to find a home. They can't just float aimlessly for the rest of their lives. So, directed by these few neurons, they begin swimming. The movement seems to strengthen the brain and the nervous system connections. The squirt may even add a few dozen brain cells while wandering. But then it finds an underwater rock, ship hull, or perhaps a lazing walrus and attaches itself. Adult squirts are sessile; they pass the rest of their lives clamped to a single surface, waving with the tides but otherwise never moving from that spot.

So their brains die. The neurons and nervous system connections shrivel and are absorbed into the squirt's soggy tissues. There is "a strong relationship between activity and brain function in animals," according to Fernando Gomez-Pinilla, Ph.D., a professor of physiological science at the University of California, Los Angeles. When the squirt stops moving, Dr. Gomez-Pinilla says, "it has no further use for a brain."

The moral of the story is that this could happen to anyone.

Sound Body, Sound Mind

Thomas Jefferson, who famously enjoyed farming and other vigorous types of seed sowing, once wrote, "A strong body makes the mind strong." He was half-right. A strong mind also makes the body strong. The connections between movement and thinking are intricate, additive, and multidirectional. Without a brain, you can't move, squirt. But without frequent movement, you have a less healthy, hardy brain.

In fact, today's most exciting research in exercise science involves deciphering the ways in which activity affects the mind, because it affects the mind in almost every way. Recent studies credibly have established that exercise stimulates the creation of new brain cells, pumps up existing ones, improves mood, aids in multitasking, blunts aging-related memory loss, sharpens decision making, dulls stress, enfeebles bullies, and if you happen to be an elementary school student, improves your math grade.

"What exercise does for thinking is remarkable," says Charles Hillman, Ph.D., a professor and director of the Neurocognitive Kinesiology Laboratory, at the University of Illinois at Urbana-Champaign. "It's effective in young people, older people, college students," college professors, and people who write or read books about exercise. "Even a small amount of activity can make an enormous difference" in the functioning of the brain, he says.

There had been hints of that possibility for thousands of years, of course. Scientists, philosophers, and mystics (who, historically, were occasionally the same person) long have talked of a mind-body connection. The first-century poet Juvenal praised *"mens sana in corpore sano"* ("a sound mind in a sound body"). But it wasn't until the past decade or so, with the advent of functional magnetic resonance imaging machines, cellular dye tagging, advanced electron microscopy, and other technologies that can zero in on the activities of individual brain cells that scientists began to understand just how a healthy body makes for a healthier brain at a molecular level.

At the same time, what the brain can do for exercise is substantial, too. The right attitude and thinking can improve athletic confidence, steady your putting, or allow you to lollop along a tightrope three stories above the floor without falling.

The poor sea squirt has no idea what it's been missing.

"None of Us Wants to Lose Our Minds"

When Canadian researchers measured the energy expenditure and cognitive functioning of a large group of elderly adults over the course of two to five years, the results were predictable. Most of the volunteers did not exercise, per se, and almost none worked out vigorously. Their activities consisted of "walking around the block, cooking, gardening, cleaning, and that sort of thing," says Laura Middleton, Ph.D., a professor at the University of Waterloo, who led the study.

But even so, the effects of this modest activity on the brain were remarkable, Dr. Middleton says. While the wholly sedentary volunteers—and there were many of these—scored significantly worse over the years on tests of cognitive function, the most active group showed little decline. About 90 percent of those with the greatest daily energy expenditure could think and remember just about as well, year after year.

"Our results indicate that vigorous exercise isn't necessary" to protect your mind, Dr. Middleton says. "I think that's exciting. It might inspire people who would be intimidated about the idea of quote-unquote 'exercising' to just get up and move."

Who, after all, wants a memory like a . . . you know . . . thing with holes in it? You use it to drain pasta? But mild cognitive decline is extremely common. This isn't Alzheimer's, but the more mundane, creeping memory loss that begins about the time our thirties recede, when car keys and people's names evaporate. So far medications have shown little promise against this insidious slide in our ability to remember and think.

But activity has a demonstrable benefit. In another recent study, a large group of women, most in their seventies, with vascular disease or multiple risk factors for developing that condition completed cognitive tests and surveys of their activities over a period of five years. Again, the women were not spry. There were no marathon runners among them. The most active walked. But there was "a decreasing rate of cognitive decline" among the active group, the authors wrote. Their ability to remember and think did still diminish, but not as rapidly as among the sedentary.

The benefits of exercise on thinking aren't limited to older adults, either. Scientists at the University of Illinois have studied school-age children and found that those who have a higher level of aerobic fitness processed information more efficiently; they were quicker on a battery of computerized flash card tests. The researchers also found that higher levels of aerobic fitness corresponded to better standardized test scores among a set of Illinois public school students.

But the impacts on aging are, for those of us in the midst of that process, probably the most beguiling. "If an inactive seventy-year-old is heading toward dementia at fifty miles per hour, by the time she's seventy-five or seventy-six, she's speeding there at seventy-five miles per hour," says Jae H. Kang, Sc.D., a professor of medicine at Brigham and Women's Hospital, at Harvard Medical School. "But the active seventy-six-year-olds in our study moved toward dementia at more like fifty miles per hour." Walking and other light activity had bought them, essentially, five years of better brainpower.

"If we can push out the onset of dementia by five, ten, or more years, that changes the dynamics of aging," says Eric Larson, Ph.D., a researcher in Seattle who studies exercise and the brain and wrote an editorial about the studies of older women. "This research is a wake-up call," he adds. "None of us wants to lose our minds," a sentiment with which I fervently agree.

You Must Remember This

What does happen to our thinking as we age? To better map the landscape of memory loss, researchers at Johns Hopkins University and the Center for the Neurobiology of Learning and Memory, at the University of California, Irvine, had groups of young and older volunteers watch pictures flash onto a screen, while the scientists watched their brains in action at the very moment that they were in the process of trying to create and store certain new memories.

Specifically, the volunteers, wearing head sensors, were shown a series of pictures of everyday objects, like computers, telephones, pineapples, pianos, and tractors, and asked to press a button indicating whether each object typically was found indoors or outside. Later they were shown another set of images and asked whether they remembered seeing that specific photo before or a similar one (a baby grand piano instead of a full grand, for instance), or whether the picture was completely new to them. The researchers tracked brain activity throughout both tasks.

There are many different types of memory processing, but one of the more important for everyday functioning is pattern separation. "Take breakfast," says Michael Yassa, Ph.D., a professor of psychological and brain sciences at Johns Hopkins, who led the study. Most of us follow a routine and eat much the same thing at the same time for breakfast most days, he says. But each morning's meal is unique and should produce a unique set of memories. "You need to be able to separate those memories and keep them apart," he explains. "Otherwise they can override one another and confuse things."

It turned out that young adults in their twenties were quite good at differentiating the images into the right category, and that activity in an area of their brain called the hippocampus increased as they did so. The hippocampus plays an enormous role in how mammals create and process memories; it also affects cognition, the basic ability to think. If your hippocampus is damaged, you most likely have trouble learning facts and forming new memories. Age is a factor, too. As we get older, our brain tends to shrink in volume, and one of the areas most prone to this shrinkage is the hippocampus. This can start depressingly early, in your thirties. Many neurologists believe that the loss of neurons in the hippocampus may be a primary cause of the normal cognitive decay associated with aging, while also contributing to disease. A number of studies have shown that people with Alzheimer's disease and other forms of serious dementia tend to have smaller-than-normal hippocampi.

In the Johns Hopkins study, the young people's hippocampi lit up with activity when they looked and mentally sorted the images. "There would be a lot of activity when young people saw either new or similar objects," Dr. Yassa says. Their brains, via the hippocampus, were learning and storing the new images as new images, even when they were quite similar to the images they had seen before.

The memories of the older volunteers, ages sixty to eighty, were not as sharp. They usually referred to pictures that were similar but not identical to ones they'd seen earlier as "old" photos. Their brains didn't create a completely new memory to correspond to the slightly different picture. The baby grand didn't register as different from the full grand. Meanwhile, their hippocampi showed far less activation than the young people's.

At the same time, Dr. Yassa says, in a separate part of his experiment, he used sophisticated MRI scanning technology to examine the interconnections among different parts of the brain. In the process, he found that the hippocampus in many of the older, inactive volunteers was not connected as robustly to the rest of the brain as in young people. Messages stumbled on their way from elsewhere in the brain to the hippocampal memory center, and vice versa.

The older people's processing miscues weren't severe. They were small lapses. But they presumably would accumulate, becoming an impasto of forgotten moments, one breakfast fading into another and some small portion of each day being lost.

But there is hope, Dr. Yassa says. "Exercise is one of the few things that might directly change this process."

"We Knew the Brain Controls Behavior, Not That Behavior Controls the Brain"

The Morris water maze is the rodent equivalent of an IQ test: Mice are placed in a tank filled with water dyed an opaque color. Beneath a small area of the surface is a platform, which the mice can't see. Despite what you've heard about rodents and sinking ships, mice hate water; those that blunder upon the platform climb onto it immediately. Scientists have long agreed that a mouse's spatial memory can be inferred by how quickly the animal finds its way in subsequent dunkings. A "smart" mouse remembers the platform and swims right to it.

In the late 1990s, one group of mice at the Salk Institute for Biological Studies, near San Diego, blew away the others in the Morris maze. The only difference between the smart mice and those that floundered was exercise. The brainy mice had running wheels in their cages, and the others didn't.

At the time, mainstream scientists believed that the mammalian brain was a relatively rigid, inflexible organ, isolated from the physiological operations of the rest of the body behind the skull and the blood-brain barrier, which prevents the passage of large molecules into the brain. It was believed that the brain did not change much structurally over a person's life span. It couldn't. It supposedly had no ability to make new cells. In high school biology classes, most of us were taught that we had been born with a certain number of brain cells and would have only those and no other neurons for the rest of our lives. When some of this limited supply of cells died due to age or a regrettable overindulgence in beer, mental function would decline. The damage couldn't be staved off or repaired.

But under the direction of Fred "Rusty" Gage, Ph.D., a world-renowned professor in the Department of Genetics, and his colleagues, these mice proved otherwise. Before being euthanized, the animals had been injected with a chemical compound that incorporates itself into actively dividing cells. During autopsy, those cells could be identified by using a special dye. Gage and his team presumed they wouldn't find such cells in the mice's brain tissue, but to their astonishment, they did. Up until the point of death, the mice had been creating fresh neurons. Their brains were regenerating themselves.

All of the mice showed this vivid proof of what's known as neurogenesis, or the creation of new neurons. But the brains of the athletic mice showed much more. These mice, the ones that had scampered on running wheels, were producing two to three times as many new neurons as the mice that hadn't exercised.

But does neurogenesis also happen in the human brain? To find out, Dr. Gage and his colleagues obtained brain tissue from deceased cancer patients who had donated their bodies to research. While still living, these people had been injected with the same type of compound used on Dr. Gage's mice. (Pathologists were hoping to learn more about how quickly the patients' tumor cells were growing.) When Gage dyed their brain samples, he again saw new neurons. Like the mice, the humans showed evidence of neurogenesis, and this neurogenesis was centered almost exclusively in the hippocampus.

Dr. Gage's discovery hit the world of neurological research like a thunderclap. Since then, scientists have been finding more evidence that the human brain is not only capable of renewing itself but that exercise speeds the process. "We've always known that our brains control our behavior," Dr. Gage told me, "but not that our behavior could control and change the structure of our brains."

The human brain is extremely difficult to study, however, especially when a person is still alive. Without euthanizing their subjects, the closest that researchers can get to seeing what goes on in the skull is through a functional MRI machine, which measures the size and shape of the brain and, unlike a standard MRI machine, tracks blood flow and electrical activity.

Not long after Dr. Gage and his colleagues published their seminal studies of neurogenesis in mice and humans, neuroscientists at Columbia University, in New York City, set out to determine if something similar was happening in living humans. They gathered a group of men and women ranging in age from twenty-one to forty-five and asked them to begin working out for one hour four times a week. After twelve weeks, the test subjects, predictably, were more fit. Their VO2 max had risen significantly.

But something else happened as a result of all those workouts: Blood flowed at a much higher volume to the hippocampus, a part of the brain where neurogenesis occurs. Functional MRIs showed that a portion of each person's hippocampus now received almost twice the blood volume it had before. Scientists suspect that the blood pumping into that part of the brain was helping to produce fresh neurons there.

The Columbia study suggests that shrinkage of the hippocampus, so common as we age, could be slowed via exercise. The volunteers in this study showed significant improvements in their memory, as measured by a word-recall test, after they'd been working out for three months. And moreover, those with the biggest increases in VO2 max had the best scores on the test of all the participants.

"It's reasonable to infer that neurogenesis was happening in the people's hippocampi," says the leader of the study, Scott A. Small, M.D., a professor of neurology at Columbia, "and that working out was driving the neurogenesis."

Fighting Back the Shadows

Mice, like people, tend to lose their grasp on memory and clear thinking as they age. They are not intellectual giants to begin with. Young mice devote most of their brain capacity to finding food or sex. But there is a poignant downslope over time in their ability to figure out how to get the chow or the girl. They become confused and distracted. Their memories slip away like shadows.

Unless they run.

In experiments that reinforced and expanded our understanding of how moving affects thinking, scientists with the Laboratory of Neuroscience at the National Institute on Aging separated young lab mice and a similar group of elderly rodents into two groups. Half of the young and the old mice were given running wheels in their cages. The other half remained sedentary. Most mice enjoy running, and the youngsters given running wheels scampered on theirs for hours at a time. Even the elderly mice managed at least an hour a day.

Weeks passed. The mice ran or, for those without running wheels, lounged. Then each of the mice was placed in an individual Plexiglas box that included a mouse-sized light-up touch screen. Images could be flashed on the screen, which, thanks to infrared sensors, recognized the lightest of nose pokes from the mice.

They were taught with a food reward to nose flashing squares. Eventually the mice had to remember and differentiate between several squares appearing on the screen, sometimes touching and sometimes widely spaced. This tests pattern separation and other elements of mouse learning and memory.

The young running mice proved masterful. They processed the information faster and with fewer errors than the young sedentary mice. Upon examination of their brain tissues, they also turned out to have more than twice as many new brain cells in their hippocampi as the unmoving animals.

Improvements took longer and were less striking among the older exercising mice and, in fact, did not occur at all among the elderly sedentary mice. None of them ever managed to understand what they were supposed to do. But doggedly, the ancient runners started differentiating one well-separated and lit-up square from the others, earning their kibble. Unlike the inactive old mice, they were able to remember and learn.

Their brains showed little evidence of neurogenesis, though, suggesting that other processes within the brain may also be at play when we exercise.

Use Your Noggin

Different scientists have varying pet theories about how exercise prompts the brain to remodel itself, each of them involving an alphabet soup of interrelated biochemical processes. One popular hypothesis points to insulin-like growth factor 1, a protein that circulates in the blood and is produced in greater amounts in response to exercise. IGF1 has trouble entering the brain—it usually stops at the blood-brain barrier—but exercise is thought to help it to pass through the barrier, sparking neurogenesis and other changes in the brain's tissues.

Other researchers credit BDNF, or brain-derived neurotrophic factor, for many of the beneficial mental impacts of exercise. BDNF is a protein produced in the brain and elsewhere in the body. Pumped out in greater profusion during and after exercise, it's known to help neurons develop and thrive. It also allows the brain to consolidate short-term memories into long-term ones.

And then there's BMP, bone morphogenetic protein. At Northwestern University's Feinberg School of Medicine, scientists have been manipulating the levels of this protein in the brains of laboratory mice. BMP, which is found in tissues throughout the body, affects cellular development in various ways, some of them undesirable.

In the brain, BMP has been found to contribute to the control of stem cell divisions. Your brain, you will be pleased to learn, is packed with adult stem cells, which, given the right impetus, divide and differentiate into either additional stem cells or young neurons. As we age, these stem cells tend to become less responsive. They don't divide as readily and can slump into a kind of cellular sleep. It's BMP that acts as the sleep aid, says Dr. John A. Kessler, the chairman of neurology at Northwestern and an author of many studies about the substance. The more active BMP and its various signals are in your brain, the more inactive your stem cells become and the less neurogenesis your brain undergoes. Your brain grows slower, less nimble, and, no matter what your chronological age, physiologically older.

But exercise countermands some of the numbing effects of BMP, Dr. Kessler says. In work at his lab, mice given access to running wheels had about 50 percent less BMP-related brain activity than sedentary controls within a week. They also showed a notable increase in Noggin, a beautifully named brain protein that acts as a BMP antagonist. The more Noggin in your brain, the less BMP activity and the more stem cell divisions and neurogenesis in your brain. Mice at Northwestern whose brains were infused directly with large doses of Noggin became, Dr. Kessler says, "little mouse geniuses, if there is such a thing." They aced the mazes and other tests.

Whether exercise directly reduces BMP activity or increases production of Noggin isn't yet known and may not matter. The results speak for themselves. Through a complex interplay with Noggin and BMP, physical activity helps to ensure that neuronal stem cells stay lively and new brain cells are born. "If ever exercise enthusiasts wanted a rationale for what they're doing, this should be it," Dr. Kessler says.

But wait, there's more. Exercise also shapes up individual brain cells, just as it strengthens muscles. Muscles of course grow fitter if we work out, a process due in part to an increase in the number of muscle mitochondria, those tiny organelles that float around a cell's nucleus and help to create energy. The greater the mitochondrial density in a cell, the greater its vitality.

Like muscles, the brain gets a physiological workout during exercise. "The brain has to work hard to keep the muscles moving" and all of the bodily systems in sync, says J. Mark Davis, Ph.D., a professor of exercise science at the Arnold School of Public Health, at the University of South Carolina. Scans have shown that metabolic activity in many parts of the brain surges during workouts, but it was unclear whether those straining brain cells were adapting and changing as muscle cells do.

Then Dr. Davis and his colleagues let some mice run for eight weeks, while others stayed inactive. At the end of the two months, the researchers had both groups run to exhaustion on treadmills. The running mice were in better shape, lasting on the treadmills almost twice as long as the unexercised animals. Their brain cells were in better shape, too. When the scientists examined tissue samples from the exercised animals' brains, they found markers indicating substantial new mitochondrial development in their brain cells. There was nothing comparable going on in the brains of the sedentary mice.

The implications of that finding are exciting. Reenergized brain cells should be resistant to fatigue, Dr. Davis says. Since bodily fatigue is partially mediated by signals from the brain, exercising your body could be training your brain to allow you to exercise more, amplifying the benefits, which is nice of it. Revitalized brain cells also could reduce mental fatigue and sharpen your thinking, "even when you're not exercising," Dr. Davis says.

Perhaps most important, the additional mitochondrial density could, at least in theory, protect against some neurological diseases. "There is evidence [from other studies] that mitochondrial deficits in the brain may play a role in the development of neurodegenerative diseases," including Alzheimer's or Parkinson's, Dr. Davis says. "Having a larger reservoir of mitochondria" in your brain cells could provide some buffer against those conditions, he says.

"There is no medicine or other intervention that appears to be nearly as effective as exercise" in maintaining or even bumping up a person's cognitive capabilities, Dr. Hillman says.

The impacts extend even beyond the ability to think and remember. Exercise also dramatically alters how you feel.

Buddha Brain

Researchers at Princeton University recently made the remarkable discovery that the brain cells that sprout as a result of exercise seem to be preternaturally calm. In the experiment, scientists allowed one group of rats to run. Another set of rodents didn't exercise. Then all of the rats swam in cold water, which, you'll remember, they dislike. It causes stress, similar to our work deadlines or marital strife.

Afterward, the scientists examined the animals' brains. They used cellular markers to determine which of the neurons were the youngest, suggesting that they had been created in the weeks since the experiment began. They also looked for gene activity indicating that individual brain cells had responded to the stress by firing.

They found that the stress of the swimming had activated neurons in all of the animals' brains, whether they'd exercised or not. But the newborn brain cells in the running rats (which were the cells scientists assumed had been created by the running) were much less likely to express the genes indicating that they'd been active. They remained quiet. The "cells born from running," the researchers concluded, appeared to have been "specifically buffered from exposure to a stressful experience." The rats had created, through running, a brain that was biochemically and molecularly calm.

For those of us now worried about the state of our memories, word that exercise also improves mood and lessens anxiety could hardly be more opportune. And the impacts on emotion and mood are wide-ranging. In an experiment at Yale University, researchers found that prolonged exercise altered the expression of almost three dozen genes associated with mood in the brains of laboratory mice, and a study from Germany concluded that light-duty activity such as walking or gardening made participants "happy," in the estimation of the scientists. Similarly, an extremely similar experiment by scientists from Oklahoma State University found that female rats allowed to run at a moderate pace for ten to sixty minutes several times a week—my exercise regimen, in fact—behaved with robust mental health in stress tests. My husband would be surprised by that finding.

Even anger seems to yield to or moderate with exercise. In a study presented at a recent American College of Sports Medicine conference, hundreds of undergraduates at the University of Georgia filled out questionnaires about their moods. From that group, researchers chose sixteen young men with "high trait anger" or, in less technical terms, a short fuse. They were, their questionnaires indicated, habitually touchy.

The researchers invited the men to a lab and had them fill out a survey about their moods at that moment. During the two days of the study, the men were each fitted with high-tech hairnets containing multiple sensors that could read electrical activity in the brain. Next, researchers flashed a series of slides across viewing screens set up in front of each young man. The slides were intended to induce anger. They depicted upsetting events such as Ku Klux Klan rallies and children under fire from soldiers, interspersed with more pleasant images. Electrical activity in the men's brains indicated that they were growing angry during the display. For confirmation, they described to researchers how angry they felt, using a numerical scale from 0 to 9.

On alternate days, the men either sat quietly or rode a stationary bike for thirty minutes at a moderate pace while their brain patterns and verbal estimations of anger were recorded. Then they watched the slide show again.

The results showed that when the volunteers hadn't exercised, their second viewing of the slides aroused significantly more anger than the first. After exercise, however, the men's anger plateaued. They still became upset—exercise didn't inure them to the slides—but it helped them to hold their anger in check.

"Exercise, even a single bout of it, can have a robust prophylactic effect" against the buildup of anger, says Nathaniel Thom, Ph.D., a stress physiologist who conducted the study. "So if you know that you're going to be entering into a situation that is likely to make you angry, go for a run first."

Don't Let the Bullies Get You Down

Exercise also provides an emotional shield if you're heading into a situation in which the other guy has not gotten his run in and has, um, "issues," as researchers at the National Institute of Mental

Health learned when they turned mouse bullies loose on their cage mates.

In a disturbingly accurate simulacrum of many modern human office situations, researchers at the institute gathered two types of mice. Some were strong and aggressive; the others, less so. All were male. The alpha mice got private cages. Male mice in the wild are territorial loners. So when the punier mice were later slipped into the same cages as the aggressive rodents, separated only by a clear partition, the big mice acted like thugs. They employed every animal intimidation technique, and during daily five-minute periods when the partition was removed, they had to be restrained from harming the smaller mice. In the face of such treatment, the smaller animals became predictably twitchy and submissive.

After two weeks of cohabitation, many of these weaker mice were nervous wrecks. Tested in a series of stressful situations away from the cages, the mice responded with, as the scientists call it, "anxiety-like behavior." They froze or ran for dark corners. Everything upset them. "We don't use words like 'depressed' to describe the animals' condition," says Michael L. Lehmann, Ph.D., a fellow at the institute who led the study. But in effect, those mice had responded to the repeated hectoring and abuse by becoming depressed.

However, that condition didn't crop up in a separate subgroup of mice that had been allowed access to running wheels for several weeks before they were housed with the aggressive mice. These mice, although wisely submissive when confronted by the bullies, rallied nicely when away from them. They didn't freeze or cling to dark spaces in unfamiliar situations. They explored. They appeared to be, Dr. Lehmann says, "stress resistant."

"In people, we know that repeated applications of stress can lead to anxiety disorders and depression," says Dr. Lehmann. "But one of the mysteries" of mental illness "is why some people respond pathologically to stress and some seem to be stress resistant."

The answer, at least in part, may be workouts. "It looks more and more like the positive stress of exercise prepares cells and struc-tures and pathways within the brain so that they're more equipped to handle stress in other forms," says Michael Hopkins, Ph.D., a researcher affiliated with the Neurobiology of Learning and Mem-ory Laboratory, at Dartmouth University, who has been studying how exercise differently affects thinking and emotion. "It's pretty amazing, really, that you can get this translation from the realm of purely physical stresses to the realm of psychological stressors."

Of course, as we all know, mice are not people. But the scien-tists believe that this particular experiment is a fair representation of human interpersonal relations. Hierarchies, marked by bullying and resulting stress, are found among people all the time, Dr. Lehmann says. Just think of your own most dysfunctional office job. (It's also worth noting that the same experiment cannot be conducted on female mice, who like being housed together, Dr. Lehmann says, so he and his colleagues are planning to test a female-centric version, in which "cage mates are swapped out continuously," to the conster-nation and grief of the female mice left behind.)

And perhaps best of all, Dr. Lehmann does not believe that hours of daily exercise are needed or desirable to achieve emotional resilience. The mice in his lab ran only when and for as long as they wished. For his own part, Dr. Lehmann doesn't run. But he has no car and walks everywhere, and although he lives in Washington, DC, a cauldron of stress induction, he describes himself as a "pretty calm guy."

Will Any Workout Work?

Whether any one type of exercise is better than another for spurring changes in the brain remains uncertain. Most researchers are pro-ponents of endurance workouts, such as walking, running, cycling, swimming, and so on. In one of the few experiments to directly compare the effects of different types of regimens on mental functioning, twenty-one students at the University of Illinois were asked to memorize a string of letters and then pick them out from a list flashed at them. Then they were asked to do one of three things for thirty minutes—sit quietly, run on a treadmill, or lift weights— before performing the letter test again. After an additional thirty-minute cooldown, they were tested once more. On subsequent days, the students returned to try the other two options. They were noticeably quicker and more accurate on the retest after they ran compared with the other two options, and they continued to perform better when tested after the cooldown.

"There seems to be something about aerobic exercise," says Dr. Hillman, who conducted the study. "It sparks changes in the brain structure and function. It's not clear if other types of exercise can do that."

Henriette van Praag, Ph.D., an investigator in the Laboratory of Neuroscience at the National Institute on Aging, agrees. "It appears that various growth factors must be carried from the periphery of the body into the brain to start or intensify a molecular cascade there," she says. For that to happen, you may need "a fairly dramatic change in blood flow," like the one that occurs when you run or cycle or swim.

But there is some limited evidence that weight training can have beneficial impacts, as well. Recently scientists in Brazil developed the first plausible animal version of weight training. It's been difficult to study the molecular effects of resistance-style exercise on the mind because animals can't lift weights. Lab rats, mice, and other animals usually love to run and can be made to swim, so it's been easy to use animals to study aerobic exercise and the brain. (Cycling is difficult.) But lab animals can't heft barbells very well, and they don't fit on Nautilus machines.

So the Brazilian researchers had the clever notion to attach weights to the tails of a group of lab rats and have them clamber slowly up a ladder five times a week. A separate control group of rats did not exercise. After eight weeks, the tail-weighted animals had developed the bulky muscles of human gym rats, an indication that the exercise regimen was focusing on muscle. They'd also become smart, performing better than the sedentary animals on tests of memory and learning. And when the scientists compared brain tissue from both groups of animals, they found that the weight-trained rats' brains contained far more growth factors associated with neurogenesis than the sedentary animals.

The few applicable studies in humans have been encouraging, too. In one, a group of women age sixty-five or older completed twelve months of light-duty weight training twice a week. They did not do any endurance training, such as walking. At the end of the year, they performed significantly better on tests of mental processing ability than a control group of women. Functional MRI scans of both groups showed that portions of the brain that control decision making and other types of thinking were more active in the weight trainers.

"We're not trying to show that lifting weights is better than aerobic-style activity" for staving off cognitive decline, says the University of British Columbia's Dr. Teresa Liu-Ambrose, the study leader. "But it does appear to be a viable option."

Lucky Thong

Meanwhile, the mental effects of exercise flow both ways. If exercising has an impact on thinking, thinking also greatly influences exercise performance, although the impacts can be shifty and fickle. Look at confidence. Some studies have found a correlation between robust self-confidence and improved performance, but others have found that athletes who are too internally assured can ignore external cues about any given day's needs; they don't pay enough attention to their opponents, the weather, their equipment, or warnings from their own tired body, which can cause failure as well.

Confidence is especially tricky when it comes to injury risk. Athletes identified by researchers as supremely confident are believed to be at higher risk for injury, because a bulletproof ego can lead to risk taking. On the other hand, low athletic self-confidence doesn't help much, either. A recent survey of high school athletes found that those who reported in the preseason that they weren't confident about how they would perform in the games and meets ahead tended to wind up hurt, especially if they were female.

But confidence is most problematic when it's yoked to self-deception, as a fascinating study of clowns and tumblers makes clear. In the study, forty-seven athletes who were hoping to land a spot in a Cirque du Soleil show filled out questionnaires about their health and attitudes at the start of a training camp. Each of the athletes previously had been an elite competitor in gymnastics, trampoline, swimming, or diving. They were used to knowing what their bodies could do.

But they were being thrust into an entirely new discipline. The training camp regimen was strenuous and mentally trying, says Madeleine Hallé, Ph.D., a senior performance psychologist with Cirque du Soleil. The athletes were beginners again after years of being among the best in the world.

More than half of them wound up injured during the four-month camp. Some hurt themselves multiple times.

Injuries were most common among those who, according to their questionnaires, possessed low "self-efficacy," a kind of enhanced self-confidence, or the feeling that you are easily capable of performing the task ahead. But not all athletes with low self-efficacy got hurt. Some considered themselves incapable of the demands of Cirque, but they were wrong. And some who scored high on self-efficacy sustained multiple injuries. They demonstrably were not physically ready, but they thought they were.

That distinction between lacking self-confidence when you should have it and having it when you perhaps shouldn't is likely to make a big difference in performance and injury risk, says Ian Shrier, M.D., a professor in the department of family medicine at McGill University, who studied the Cirque performers. It also will affect the proper response. If you're correct that you're not physically ready to perform a task, the best intervention is going to be augmented coaching and physical training. But if you have the ability but simply don't believe that you do, intercessions should probably focus on building psychological coping skills, rather than physical technique.

Which carries us, inexorably, to the issue of lucky underwear. If you lack confidence despite having fine technique and training, science suggests that you might want to look to luck. An astonishing number of the world's top athletes are deeply superstitious. For years, Michael Jordan wore the shorts from his national-championship-winning University of North Carolina days under his Chicago Bulls uniform. Serena Williams supposedly wouldn't change her socks at tournaments she was winning. Other professional athletes carry lucky charms or perform rituals, like bouncing a basketball in elaborate sequences before a free throw or kissing the golf ball before a putt. Baseball first baseman Jason Giambi has said that he would slip on a pair of "lucky" thong underwear when his batting average fell. (And during Giambi's career the thong's reputation became so potent that slumping teammates reportedly begged to borrow it.)

But does lucky underwear work? Unfortunately for those of us who really despise thongs, researchers at the University of Cologne, in Germany, found that the answer is a qualified yes.

In a series of experiments, the scientists asked college students to make as many golf putts as possible on a putting green. Before his or her first attempt, each participant was handed a golf ball. Some were told, "Here is your ball; so far it has turned out to be a lucky ball." The rest were told, more blandly, "This is the ball everyone has used so far." Each student putted ten times.

The students using the "lucky" balls sank significantly more putts than those who didn't.

Next the researchers had a different group of students complete a dexterity test. The students were given a plastic cube containing thirty-six balls and a shelf dimpled with thirty-six holes. They were told to dip and twist the box until the balls rested in the holes. First, though, they were given instruction from a moderator, who told some of the volunteers, "I press the thumbs for you," a German idiom that loosely translates to, "I'm keeping my fingers crossed for you; good luck." The rest received neutral directions. By a fairly significant margin, the volunteers who had been offered the thumb pressing maneuvered the balls into position fastest.

"Activating a good-luck superstition," the authors concluded, "leads to improved performance by boosting people's belief in their ability to master a task." More precisely, they added, "the present findings suggest that it may have been the well-balanced combination of existing talent, hard training and good-luck underwear that made Michael Jordan perform as well as he did."

Interestingly, superstitions flourish most in situations in which talent is being pushed to its limits and any edge might be decisive, even if it's fantastical. In a nifty experiment at Colorado College, in Colorado Springs, a group of students putted. Their first round of putts was easy, measuring only three feet to the cup. The second round consisted of nine-foot putts. Each volunteer putted twenty times at each distance. Students could choose their balls from a basket containing four different colors. During the easy round, the best putters pulled balls out at random; they weren't interested in the colors. But the less able students, those who weren't good at putting, tended to pick the same-color ball after any successful putt; it had become their "lucky" ball.

When the testing moved to the longer putts, the better golfers started picking the same-color ball after successful putts. As their skills were being challenged, they began turning to luck to increase their chances. Meanwhile, the less talented putters, who missed almost all of the longer putts anyway, no longer seemed to care which ball they used. Luck couldn't help them now.

The lesson from this and the other experiments is, at its most basic, that being superstitious is a sign not of weakness but "of hope," says Kristi Erdal, Ph.D., a professor of psychology at Colorado College and an author of the putting study. You may be turning to an external, intangible force, but you haven't given up.

And that, in a broader sense, is the message of all of the science related to exercise and the brain: Just keep going. Every researcher I spoke with on this topic exercises. Some run. Some walk. There are a few bike racers. Tennis is popular, too. But none are sedentary. They know too much. Beneficently, they'll share.

"As a neurologist," Columbia University's Dr. Small says, "I constantly get asked at cocktail parties what someone can do to protect their mental functioning. I tell them, 'Put down that glass and go for a run.'"

How to Sharpen Your Mind and Mood

1. Bulk Up Your Brain.

In one study, elderly sedentary people who began a walking program showed significant growth in several areas of the brain after six months. Scientists believe that the workouts prompted the creation of new neurons, as well as new blood vessels and connections between the neurons. The walkers' brains were bigger, faster, and younger, and they consequently performed better on tests of memory and decision making than people who'd remained sedentary.

2. A Little May Be Enough.

In mice, a fairly short period of exercise and a short distance seems to produce results in terms of improved cognition. "Walking around the block, cooking, gardening, cleaning, and that sort of thing" significantly improved cognitive function in a group of older people, says Dr. Middleton, who studied the group.

3. Run Away from Serious Memory Loss.

"Epidemiological studies show that long-term runners have a lower risk of neurological disease," including Alzheimer's and Parkinson's, says Dr. Mark Tarnopolsky, a professor of medicine at McMaster Children's Hospital, who has studied exercise and the brain for decades and himself runs almost every day.

4. Get the Kids Out, Too, for Your Own Sake.

Studies from the University of Illinois have found that "just twenty minutes of walking" before a test raises kids' scores, even if the children are otherwise unfit or overweight, says Dr. Hillman, who has studied exercise and kids' brains. Other work from his lab has shown that aerobically fit children score higher on tests of complex memory than less fit youngsters. But perhaps most compelling from a parental standpoint, a years-long Swedish study found that among more than a million eighteen-year-old boys who joined the army, better fitness consistently correlated with higher IQ, even among identical twins. The fittest young Swedes also were significantly more likely to go on to lucrative careers than the least fit, rendering them less prone, one hopes, to taking up residence in their parents' basements.

5. Quit with the Dumb Jock Jokes.

Japanese researchers recently loaded rats' running wheels, similar to cranking up a stationary bicycle's resistance. The load on the wheels equaled 30 percent of the rats' body weight. They could barely push themselves along, straining as if against a hurricane-level headwind. After eight weeks, the animals had packed on muscle mass in their legs, while a group of rats jogging easily on unloaded wheels had not. The buff rats also displayed increased levels of gene activity in the brain associated with improved brain functioning—more, in fact, than in the animals that hadn't added muscle. The stronger the animals became, the better their brains worked.

6. Take a Step. Lift a Mood.

Exercise speeds the brain's production of serotonin. Abnormally low levels of serotonin have been associated with anxiety and depression. In some studies, exercise has been as or even more effective than antidepressant medications at making people feel better.

7. Be Patient.

The stress-reducing changes in the brain wrought by exercise don't happen overnight. In experiments at the University of Colorado, rats that ran for only three weeks did not show much reduction in stress-induced anxiety, but those that ran for at least six weeks did. "Something happened between three and six weeks," says Dr. Benjamin Greenwood, who helped conduct the experiments. The lesson, he continues, is "don't quit." You may not feel a magical reduction of stress after your first jog or swim. But the molecular biochemical changes will begin, Dr. Greenwood says. And eventually, he says, they become profound.

8. Find a Training Partner.

A rather touching experiment with a species of sociable, gregarious rats found that when these animals were housed alone, their brains did not benefit from exercise as much as when they were in shared cages. Loneliness increased the levels of stress hormones in the animals' brains. Exercise added more stress, apparently blunting the positive effects of the workouts. Socially housed rats produced copious amounts of new brain cells when they exercised; the lonely animals did not.

9. Get Horizontal.

Sex can spur neurogenesis. It is a moderate workout, after all, if you do it right. When male rats at the Princeton Neuroscience Institute were given access to "sexually receptive" females, they responded as nature intended and vigorously engaged with the girl rats. The resultant activity led to an increase in neurogenesis in their brains. Sex improved their ability to think, obvious jokes notwithstanding.

9

Survival of the Fittest

If there had been no *Australopithecus afarensis*, we might not have such a need for the Stairmaster. One of modern man's more enterprising ancestors, *A. afarensis* stood up on two legs about three and a half million years ago. Other humanoid species had become bipedal before him. But *A. afarensis* was better at it, and graceful bipedalism changed everything for humanoids. These early men could see farther than other animals, since their eyes were now situated higher up and able to see above the top of the rippling savannah grasses. With a vertically oriented chest, their lungs could expand more; they could draw in deeper breaths.

And most important, they could move differently. Some time ago, anthropologists and other scientists established that the most efficient stride for distance travel is walking and the most efficient posture for walking is upright. They determined this by the ingenious expedient of filming a variety of animals—antelope, donkeys, dogs, and men—while they sprinted, cantered, galloped, or strolled on treadmills and also testing their oxygen consumption. Bipedal men turned out to be second-rate runners compared to quadrupeds. Four-legged animals run faster than we can and use much less oxygen to go the same distance. But we can out-walk them. Bipedal human walking is quite efficient. A prehistoric humanoid would have been able to walk a mile and use much less energy than an antelope required to sprint that same mile. So Urk, though slow, would have been implacable, stalking the speedier antelope until it simply keeled over and provided refreshments for everyone.

These hunts shaped the futures of both species, stalkers and prey. "Locomotion, movement through the environment is the behavior that most dictates the morphology and physiology of animals," wrote biologists from the Department of Integrative Biology at the University of California, Berkeley, in a recent issue of *Science* magazine.

Essentially, the key to early man's success in his environmental niche was his ability to move as he did. "Activity was obligatory for survival," writes the University of Missouri's Dr. Frank Booth, an expert on human physiology and inactivity. Early men who were out and about all day acquired more meat and mates than the layabouts. They sired more children. Their genetic traits survived and accreted. Those "not able to perform physical activity," Dr. Booth writes, were doomed "to have their gene pools selected for extinction."

Over the millennia, the imperative to move intensified and became increasingly interwoven with human-ness. By the Paleolithic age, beginning some eighty thousand years ago, humans could have been defined as a people in motion. The surviving human species, *Homo sapiens*, had lithe bodies and large brains that required frequent feeding (brains are very metabolically active). These larger-scale thinking organs directed them to hunt and gather efficiently, giving *Homo sapiens* the additional meat necessary to allow their brains to grow even larger. Complex thinking both resulted from and drove movement.

Humans "evolved to feed, shelter, and invent while ambulatory," says James Levine, M.D., a physician and researcher at the Mayo Clinic in Rochester, Minnesota, who has studied the physiology of inactivity for years.

Physical activity seems to be "an evolutionarily programmed necessity in our genes," writes UCLA professor Dr. Fernando Gomez-Pinilla in a wonderfully titled academic article "Revenge of the 'Sit.'"

By the end of the Paleolithic age, about ten thousand years ago, men were men (or women). The fundamental human genome was in place.

It has changed surprisingly little since. "About ninety-five percent" of the DNA that each of us carries within us is identical to that carried by cavemen, according to Dr. Booth. So while we still have "Stone Ages' genes," he says, we live now in "space-age circumstances." It's an uneasy mix.

A SNP Here, a SNP There

This chapter is out of date. It would have been out of date if you'd read it the week I finished it. There are few scientific fields changing as rapidly as those that involve genetics, and this is especially true of exercise-science genetics. It wasn't until the past decade that physiologists and other researchers began using modern gene-sequencing techniques to look for specific portions of the human genome that might influence health and fitness or athletic performance. In 2000, an exercise-related "map" of the human genome, compiled and published by researchers from the Pennington Biomedical Research Center and other institutions, highlighted twenty-nine "genes and markers" from the full human genome for which scientists had "evidence of association or linkage with a performance or fitness phenotype in sedentary or active people." The most recent version of a similar gene map, published in 2011, included data on more than eight hundred genes. Others have been identified since, probably as recently as this morning.

But the effort is Sisyphean. The human genome is stunningly

complex, and so is human movement. According to the results of the Human Genome Project, our DNA contains around twenty-five thousand genes, which is relatively few—actually, it's only a factor of two greater than the number of genes possessed by the round-worm, which should serve to keep us humans humble.

But individual genes are only the most obvious element in DNA. Each gene contains millions of bases, composing the nucleo-tides. As you have forgotten from biology class, nucleotides are the two twining units that create the distinctive double helix; they are themselves made up of four different repeating molecules (adenine, cytosine, guanine, and thymine), the bases. There are three billion or so sequences of these four bases in the human genome, and their order is not precisely the same from person to person. Each of us is endowed with particular SNPs (single nucleotide polymorphisms), a variation within a single sequence of the bases—a cytosine occur-ring in one person where a thymine is in another, for example. By most estimates, the human genome contains hundreds of millions of different SNPs.

It's largely because of SNPs that each of us varies from one another and, in broader terms, from Paleolithic men. The vast ma-jority of our genome is shared and unchanged from ten thousand years ago, but there has been shifting and rearrangement over the millennia. Countless SNPs have come and gone, with whatever re-action that they caused in the body not conferring any evolutionary advantage and thus disappearing.

Genes do not operate in a vacuum, after all. They work inside the body. Gene expression, or gene activity, drives biology. This ac-tivity is controlled by the gene's many individual SNPs, translating the information encoded in the DNA into a particular protein, which then affects specific bodily systems in one way or another. We just don't necessarily know what those ways are or how they inter-relate to create, say, a Michael Jordan and his ability to levitate from a basketball court while the rest of us are leaden footed.

"The genome is a vast landscape of greater than three times ten to the ninth power letters," or bases, "the sequence of which is known but the function of which is still largely hidden from our understanding," according to Stephen Roth, Ph.D., a professor at the University of Maryland and one of the world's experts on exercise genetics.

This mystery is really what makes exercise genetics so absorbing and worth the attention of those of us who are otherwise apathetic about guanine drift. Exercise genetics is the story of what makes us able to move and to move more or less well. The discoveries are ongoing and sometimes confusing, but the latest findings in exercise genetics do provide tantalizing clues about why some people are so much faster than I am, as well as why slow people like me still feel an inchoate drive to keep going. It's also about whether and why there are physiological barriers to fitness for some people, as well as how there's a possibility to tweak your child's genome before she's even born, increasing the chances that she'll become a professional athlete and support you with her signing bonus.

Is It Your Fault That You Don't Want to Work Out?

Not long ago, a consortium of European researchers delved into the activity habits of 37,051 sets of twins. As you might have guessed, twins are eternally popular with geneticists, because they provide a neat statistical model for determining whether a behavior is influenced by genetics or almost exclusively by environment. Identical twins share 100 percent of their genome; fraternal twins share 50 percent. All twin pairs, if raised together, share approximately the same early environment. So if a behavior is more common between identical twins than between fraternal twins, it is presumably being directed to some degree by genes.

In this particular study, scientists looked at the simple decision to exercise or not. They turned to survey data covering twin pairs aged nineteen to forty in Australia, Denmark, Finland, the Netherlands, Norway, Sweden, and the United Kingdom. Although the researchers set an exceedingly low standard of one hour per week of light jogging or a similar activity to classify someone as an "exerciser," only about 44 percent of the males and 35 percent of the females met the standard. Across the board, however, the identical-twin pairs were more likely to share exercise habits than the fraternal twins.

Using complicated statistical formulas, the scientists determined that differences in exercise behavior were about 60 percent attributable to genes. In other words, your parents continue to exert considerable influence over your decision about whether to be active, not just by signing you up for soccer camp when you're a kid but also by bequeathing you a genetic urge to work out—or not.

"Most people probably hadn't thought of exercise behavior as a domain in which genetics would be involved," says Tuomo Rankinen, Ph.D., a professor with the Human Genomics Laboratory, at the Pennington Biomedical Research Center, and an authority on exercise genetics.

Other researchers immediately turned to replicating and expanding the work. The resulting studies have, almost without exception, reinforced the idea that the drive to exercise or avoid exertion is inherited to some extent. In one notable animal experiment, mice were bred to run voluntarily for hours. Since they had the choice not to run—wheels were placed in their cages, but they were not forced onto them—presumably they enjoyed the activity, if we can anthropomorphize enjoyment to mice. Examining the genomes of these avid runners, researchers at the University of North Carolina at Chapel Hill found thirty-two points in the DNA that seemed to have a "significant" association with the desire to run and thirteen that were "suggestive." Mice that displayed less interest in running rarely shared these genetic variants. The runners also were more confident in mazes and otherwise seemed behaviorally distinct from the other animals. "These results reinforce a genetic basis for the predisposition to engage in voluntary exercise," the authors concluded, while gamely adding that the "many genomic elements" that "contribute to the predisposition for voluntary exercise" are still "not yet well understood."

That point is amplified by a study published recently in *Medicine & Science in Sports & Exercise* from scientists working at the University of Missouri, in Kansas City. They sequenced portions of the genomes of humans who readily exercise and those who avoid exercise at all costs. Using blood samples from more than twenty-six hundred people, the researchers combed through more than a million and a half sites along each person's DNA. They found that people who were active (but not necessarily athletic) tended to have similar SNP variations on several different genes. The genes in question didn't affect obvious physical characteristics, like speed and strength. Instead, the genetic differences were subtle. One of the affected genes is thought to influence how people respond to fatigue, suggesting that for some people the same amount of exercise may be more tiring—and therefore less appealing—than for others, even if they are equally fit. Another gene is widely expressed in both muscles and the brain and is likely to have an impact on how physically easy and mentally rewarding exercise feels. Yet another gene has been linked to how well the body regulates energy, which can have an effect on the desire to exercise.

These findings reinforce the complexity of exercise genetics. There's no one gene for "liking to exercise." There are instead a constellation of genes, with innumerable SNP variations and other elements of DNA that influence certain bodily functions that relate to how and whether we exercise. A similar complexity rules all aspects of movement and genes. There is no one gene for fast marathon running or for a world-class tennis serve. As Richard

Dawkins, the great Oxford biologist and professional curmudgeon, has written, there is "no gene 'for' anything." There are genes that, in expressing a certain protein, switch on bodily systems that affect hormone levels or metabolism or other cellular activity in the body, which in turn initiate or halt biochemical processes that build bone or cause muscle to waste or prompt the creation of new brain cells. Those bodily changes then improve your ability to exercise, or they make workouts difficult—and which of those two outcomes prevails may depend on other changes being caused simultaneously by the expression of other genes that affect other parts of your body.

Just recently, for instance, scientists identified a specific variation of one gene that seems to influence the cellular makeup of collagen, one of the building blocks of connective tissue. The composition of your collagen affects the tensile strength of your ligaments and tendons and joint range of motion, which then has an impact on athletic ability and injury risk. Women with the one particular variation of the gene seem to be at heightened risk for an ACL tear. But both men and women with the same gene variant may have improved running ability, since their tendons are likely to be stiff and return more energy with every running stride.

But that potential physiological advantage may manifest itself only—and I'm speculating here—if the same person possesses another recently identified genetic variant that is believed to create the right internal conditions for optimism. That variant was found to be more prevalent in people who have a buoyant sense of possibilities, who think that goals are attainable and they will attain them. Like the little train, they think they can.

And then, of course, there's your environment. The interactions between it and your genome become significant early in your lifetime. Quite early.

The Heart of a Runner

Almost anyone who's been pregnant remembers the profound link that develops between a mother-to-be and her unborn child. You feel that life inside you, both physically independent and braided with your own. The bond runs deeper, though, than most of us might have imagined. A baby's time in the womb can change his or her physiology and genome. It can shape, to some extent, his or her future health and athletic hopes.

Not that I want to put pressure on any of you out there who are pregnant, of course.

An interesting recent essay in the *British Journal of Sports Medicine*, written by scientists affiliated with Greece's FAME Laboratory, at the Centre for Research and Technology, noted that prenatal circumstances affect the "malleability of mammalian biology." In particular, they point to the case of unborn Kenyans. They are carried in the wombs of mothers who live at high altitude and who, with rare exceptions, are the products of generations of mothers who lived at the same altitude.

There is a process in human development known as epigenetics. It involves the bequeathing of traits from one generation to another without those traits being coded into the DNA. These traits are a physiological response to the environment that is robust enough that you pass the change along to your children. In mice, it's been found that fathers that eat a high-fat diet during their mouse adolescence change their metabolism in ways that eventually affect their pups' risk for diabetes, even though the dietary habits had been under-taken long before the male mice impregnated a female—and may have been changed in the years since.

Epigenetic alterations can occur within a generation, but more frequently seem to accumulate over several generations, eventually gathering enough force to show up in children's physiology. That dynamic seems to be at work among Kenyan babies to some extent, the Greek scientists suggest. Their mothers, living at altitude and having been born to mothers who did likewise, have wombs that are thought to pump more blood to the unborn child than in sea-level pregnancies, changing the infant's own vascular system. Those changes then could be expected to alter how the child's genome functions, with some genes being upregulated in response to the needs of the baby's circulatory system and others muted. Babies born outside high-altitude Kenya would be unlikely to share those same responses.

But even here in the West and at tolerable altitudes, science shows that different prenatal environments shape different babies. In an emotionally rich experiment completed recently at the Kansas City University of Medicine and Biosciences, scientists revisited a group of healthy women aged twenty to thirty-five who'd been part of a pilot study of exercise and pregnancy. About half of the women had exercised during their pregnancies, jogging, power walking, or otherwise working out moderately a few times a week. The other pregnant women "were normally active but did not engage in formal exercise," says Linda May, Ph.D., an exercise physiologist who led the study. All came in several times for readings of their and their babies' heart rhythms.

Years ago, scientists showed that a fetus's heart rate increases while its mother is exercising. But most people had thought that the response was transient, lasting only while the mother worked out. Dr. May suspected, though, that an unborn child's heart might be more permanently remodeled by a mother's workouts. Regular exercise leads to a slower heart rate and other healthy changes in most people's heart rhythm. That's part of the training response.

Unborn children, it turns out, exhibit a training response, even though their mothers are seemingly doing all the work. When Dr. May examined the fetal cardiac readings, she found that fetuses whose mothers had exercised had healthier heart rhythms than those whose mothers had not worked out.

And the changes persist. In a follow-up study, she found that a month after being born, the babies whose mothers had exercised still had healthier heart rhythms than the others, with the greatest differences among the babies born to mothers who'd exercised the most. Whether these heart alterations are permanent is not yet known, but there seems little doubt that babies with stronger hearts start life with subtly altered physiology, which, again, would change how their genes and SNPs operate and how the body responds to them.

Is There a Perfect Genome?

None of which is meant to imply that it's too late for you—that there will be no marathons in your future if your mom didn't jog while pregnant; or, perhaps worse, none for your children if you didn't waddle through your third trimester or, if you're a father, abstain from potato chips as a teenager.

Genetics is not rocket science. It's not even cosmetology, which is quite a bit more precise and reliable.

There are reasons beyond the womb environment why Kenyans excel at distance running. Many rural Kenyan youngsters run more than twelve miles a day to get to and from school. My son walks a few hundred feet to our car. Early physical training affects later physical accomplishments, in part by impacting gene operations, but even more by directly molding muscles, lungs, brain, and bones.

But the issue of what makes a fine athlete—nature or nurture— is at the heart of exercise genetics. Why are some people so beautifully fleet and coordinated and others of us ungainly? Can we practice our way to the front of the pack at our next 5K, or will no amount of effort overcome my slow person's genetic inheritance?

A famous 2007 study of elite female athletes who also happened to be twins determined that about 60 percent of the women's athletic ability was probably inherited. But the study was small. There are only so many pairs of elite twin athletes around. And the methods used to study the women's genomes were by today's standards crude.

A more recent and larger study of elite athletes came to the convincing conclusion that, yeah, certain genes can help athletes to succeed. But you still can make the podium in the Tour de France without them.

For the study, scientists in England and Spain first went through the available literature and their own past experiments to come up with a group of genes and SNPs that seemed most likely to influence success in either endurance or strength sports. The genes or portions of genes were believed to affect a wide variety of physical traits, including cardiac output, muscle fiber type (whether someone was endowed primarily with fast- or slow-twitch fibers), and fuel metabolism. The chosen SNPs were blunt in their effects, influencing obvious physical attributes. The scientists did not include genes associated with subtler traits such as balance, pain tolerance, or motivation, all of which would affect elite performance but which are very difficult to quantify in terms of genetic influence.

They chose the seven SNPs that seemed most relevant to endurance performance and created a genotype score. Someone who had all of the SNPs would score 100. Then they sequenced portions of the genomes of forty-six world-class Spanish athletes, including Olympic distance runners and a top Tour de France finisher. (He is not named, although a few minutes on Google would probably provide you with his name, if you wished.)

As a group, the elite athletes had more of the SNPs in their genome than a control group of average Spanish citizens. But none of the athletes scored 100. Only three had six of the seven relevant SNPs, and they did not happen to be the most successful athletes in the group. The Tour de France rider had only three of the SNPs, which has not seemed to impede his riding.

Repeating the test among power athletes (such as javelin throwers and weight lifters) with a different set of genes produced similar results. Most of the power athletes had some of the power-associated genes. None had all of them.

The findings reflect those from earlier, less sophisticated studies that looked at single genes. There was at one point excitement about the ACTN3 gene. Scientists at a number of different universities on several continents announced independently that a disproportionate number of world-class sprinters carried that particular gene. It was dubbed the "speed gene."

Shortly afterward, other researchers published reports that they'd found the "strength gene." Known as ACE (an acronym for angiotensin converting enzyme), it clustered in the genomes of athletes who excelled at power sports, such as shot put and weight lifting. A few lucky sprinters turned out to carry both the ACTN3 and ACE genes. They should, thanks to their DNA, have been unbeatable.

But they weren't. In recent tests, some of the world's top sprinters were found to harbor neither of those particular genes. They were muddling through somehow, though.

"There are indeed numerous other contributors to the 'complex trait' of being an athletic champion that are not likely to be reducible to defined genetic polymorphisms," the authors of the genome scoring study wrote. Those factors include "technique, kinematics [coordination], motivation, pain tolerance. Athletic success is also influenced by 'external' factors that are totally independent from genetic endowment (e.g. social support or economic possibility)." Money, encouragement, and luck play an incalculable role in athletic success.

But no element outside of genetics is as important as practice. K. Anders Ericsson, a Conradi Eminent Scholar of Psychology at Florida State University, famously pronounced that it takes ten thousand hours of practice in any endeavor to acquire expertise.

That dictum has not consistently held true in sports and exercise. The original studies from which Dr. Ericsson developed this theory involved young violin players. The best players had devoted themselves to practicing for several hours a day for ten years or more, equaling at least ten thousand hours of uninterrupted rehearsal. It's worth noting that they were not getting out into the fresh air much.

The ten-thousand-hour rule is squishier in sports and exercise. Certain requirements for success are not going to yield to any amount of practice. Ten years of training will not turn a heavy, 6'5" man into a female Olympic gymnast. (Gender is, of course, the most ineluctable genetically controlled trait in athletics.) In the same way, a 4'8" string bean is unlikely to play in the NBA or win a high-jumping medal.

But practice will go some way toward intensifying or altering whatever physical gifts or deficiencies you might have. In study after study, the primary determinant of success in distance running (which gets studied more than most sports) is training volume. Whoever practices the most tends to win or do well. Although it also helps to be Kenyan, proving that the plaiting of practice and genetics, of nature and nurture, remains fascinatingly intricate.

The Outlier Enigma

But for most of us, who don't plan to vie for Olympic gold but wouldn't mind feeling less winded after taking the stairs, the most compelling exercise-related genetics research is looking into why sometimes exercise simply doesn't take, and what that may say about the state of the human genome.

Hidden away in the results of almost any study of exercise programs is the fact that some people do not respond at all, while others respond at an unusually high rate. Averaged, the results may suggest that a certain exercise program reliably produces certain results—

that jogging, say, three times a week for a month will improve your endurance capacity or reduce blood pressure; and for almost any given group of exercisers, those results are likely to hold true. But for outliers, the impacts can be quite different.

That reality was highlighted by the results of several different new studies. In one, researchers in Finland recruited a large group of healthy but sedentary adults for a study of how different people's bodies respond to different types of exercise. Sloth and inactivity are sadly common in Finland today, as in most of the developed world, despite the region's fabled history of stomping other nations in cross-country skiing. So researchers had little trouble finding several hundred men and women between the ages of forty and sixty-seven who did not work out. They completed a stationary bicycle session to measure their current baseline aerobic fitness. They also underwent a battery of strength tests on weight machines, to see how much (or how little) muscular power they possessed.

The scientists then divided the volunteers into four groups. One was assigned to complete an endurance-training program consisting of biweekly workouts on a stationary bicycle. In the beginning, the riding was gentle and short, with riders pedaling easily for thirty minutes. But after a few weeks, the intensity, as well as length, of the rides increased, until eventually the volunteers were pedaling for up to ninety minutes at a time, with one session each week including some blood-pumping, timed sprints.

Another group began strength training. Twice a week, they showed up at a gym for supervised sessions with a trainer. Theirs was hardly a *Pumping Iron* regimen. Weights were low to start, barely half of the maximum weight that each person could lift on his or her own. Everyone completed about eight or nine upper- and lower-body exercises on weight machines. After a few weeks, the intensity was ratcheted up, with the amount of weight each person lifted and/or the number of repetitions of each exercise climbing steadily.

The third exercising group completed both endurance and

weight training sessions several times every week, while the final group of volunteers was asked to perform no exercise, to serve as a control.

The experiment lasted for twenty-one weeks, with the time and effort required of each volunteer increasing over time. Many of them ended the experiment in much better shape than when they had started. The biggest gainers had improved their fitness or strength by as much as 42 percent.

But not all of them had improved. When the entire group re-took the cycling endurance and muscular strength tests, some were at the same fitness level as they had been twenty-one weeks before, and a few performed more poorly, by as much as 18 percent. They apparently had become almost 20 percent less fit or strong than they had been before they began exercising.

The range of response was especially wide among the group that performed both endurance and strength training. Some bettered their strength noticeably but displayed no similar bump in endur-ance. Others became aerobically fitter but not stronger, while still others showed no improvements in either area. Only a fortunate few became both fitter and more buff. As the researchers wrote, there were "large individual differences . . . in the responses to both en-durance and strength training."

But why? Why wouldn't everyone benefit almost equally from exercise?

Since this is a chapter about the genetics of exercise, you can probably hazard a guess. And the first-ever large-scale look at the genomes of people who either did or did not respond physiologically to workouts would confirm it.

In that study, researchers from the United States turned to ge-netic data about 473 healthy white volunteers who were enrolled in the Heritage Family Study, a years-long examination of exercise genetics. Each of the volunteers had already completed a carefully supervised five-month exercise program, during which they pedaled stationary bicycles three times a week at identical intensities. Some wound up much fitter, as determined by the increase in their VO2 max. In others, VO2 max barely budged. No obvious, consistent differences in age, gender, body mass, or commitment marked those who responded well and those who continued to huff and struggle during their workouts even after five months.

But there was a divergence in their genomes. Using sophisticated, relatively speedy genome sequencing techniques, the researchers looked at 324,611 individual SNPs. Each of these SNPs had been identified by other researchers as coding for proteins that in one way or another might affect exercise response. The researchers wanted to see who had which SNPs.

In the end, they identified 21 SNPs, out of the more than 300,000 examined, that differed consistently between the two groups. SNPs come in pairs, since each of us receives one paternal copy and one maternal copy. So there were 42 individual versions of the 21 SNPs. Those exercisers who had 19 or more of these SNPs improved their cardiorespiratory fitness three times as much as those who had 9 or fewer.

One SNP in particular, located on a gene known as ACSL1, seemed especially potent, possibly accounting for as much as 6 percent of the difference in response among people, a huge percentage by the standards of genetics studies. This gene is known to play a role in how the body metabolizes fat, which could partly explain why it affects exercise response. People who metabolize fat well might be able to exercise longer than those who cannot. But "far more research is needed before we can say how any particular gene influences the body's response to aerobic exercise," says Claude Bouchard, Ph.D., who holds the John W. Barton Sr. Endowed Chair in Genetics and Nutrition at Pennington and was lead author of the study.

In the meantime, it's fair to ask what's going on with the human genome that could even allow for nonresponse.

Space Age Dissonance

"The model for human physical activity patterns was established not in gymnasiums, athletic fields or exercise physiology laboratories, but by natural selection acting on eons of evolutionary experience," Loren Cordain, Ph.D., an exercise physiologist and expert on Paleolithic life, wrote with his colleagues more than a decade ago. And most physiologists still agree with that: Genetically, we remain cavemen. "The basic framework for our physiological gene regulation was selected during an era of obligatory physical activity," Dr. Booth says.

But we don't get out as much anymore. Modern hunter-gatherers, whose lifestyle presumably mirrors that of our earliest ancestors, cover between twelve and twenty miles more a day than the rest of us do. We are not reaching "the necessary levels of activity for healthy gene expression," Dr. Booth warns.

Without at least the vaunted twenty minutes of walking, he says, our bodies do not function well.

The result increasingly seems to be dissonance in our DNA, according to many scientists. "Our current genome is maladapted," Dr. Booth says. "We're expressing late Paleolithic genes in a sedentary world."

Research by Dr. Booth and others shows that thousands of elements in a person's DNA change, in fact, if he or she is sedentary for even a few weeks. Many of these elements involve messages telling genes to start encoding for proteins. Inactivity can lead genes to remain quiet when they should be busy or to express proteins that function in unexpected ways.

As a telling example, Dr. Booth points to the heart. Vigorous activity stimulates the left ventricle to grow, a healthy, desirable response prompted in part by a "selective increase" in certain genetic signals to the cardiac muscle, he says, and a simultaneous decline in four other genetic markers. Those same four markers "are markedly upregulated" by inactivity, resulting in a similar cardiac enlargement, but with a profoundly different and unwelcome outcome. An inactive enlarged heart is diseased. It has cardiac myopathy, which can be fatal.

Many other diseases and conditions seem to be related to a "mismatch" between our prehistoric genetic endowment and our modern, inactive lives. Scientists regularly point to the low incidence of diabetes among today's hunter-gatherers. (Although there is something a bit imperious and out of touch about the commentators' attitude. Many of the hunter-gatherers, given any opportunity, abandon the hunt and turn with relief to fast food and television.)

This mismatch almost certainly has a bearing on some people's nonresponse to exercise. Inactivity drives changes in so many elements of the body. Significant genetic "modulation does occur during longer term metabolic adaptation in humans, such as endurance exercise (positive) and insulin resistance (negative)," writes James Timmons, Ph.D., a professor at the University of London, who has conducted numerous studies of nonresponse to exercise in people and animals. If a body no longer uses insulin normally or otherwise exhibits dysfunction in its metabolism, it's probably not going to respond well to exercise.

"Inactivity is abnormal," Dr. Gomez-Pinilla concludes.

The solution, obviously, is to get moving. But for some people, those whose bodies respond to workouts with a physiological shrug—who grow no more or even less fit—more may be needed. At some point, genetic testing may be able to ascertain whether a person who does not respond aerobically to running might react favorably to weight training sessions.

That possibility may, in fact, offer the greatest long-term benefit from the new science of exercise genetics, Dr. Rankinen says. Learning more about the genetics of active people could allow for interventions to goose everyone else. "Right now, most people don't exercise, even though we all know that, for health reasons, we should," Dr. Rankinen says. Perhaps through exercise scientists' growing knowledge of genetics "we can find ways to help make exercise easier or more attractive for people." If, for instance, it turns out that some people have a genetic predisposition to develop especially sore muscles after running, he says, "Maybe we could start directing those people to other kinds of exercise."

But those days are still well in the future. Scientists have barely begun to untangle the many ways in which different genes influence the body's ability to move. For now, Dr. Bouchard says, "there are countless other benefits provided by exercise," regardless of whether your body responds with increased cardiovascular fitness. "Exercise can reduce blood pressure and improve lipid profiles," he says. It can better your health, even if, by certain measures, it does not render you more aerobically fit.

And many of us, once we start, find that we enjoy moving and develop a certain longing to be out, even if our parents did not bequeath us the requisite enjoy-exercise gene. Perhaps it's an atavistic echo from *A. afarensis*. But more likely it's our innate good sense. "Even at the highest percentages of likely heritability" of exercise behavior, Dr. Rankinen says, in the end "the choice to exercise is yours."

What Exercise Genetics Teaches Us

1. Move.

Our genome was largely selected in the Stone Age, when physical activity was, as one scientist says, "obligatory." Early men and women who didn't move were eaten or starved. Their genome was extinguished. Those who survived passed along genes that promote and thrive during movement. The difference between absolute inactivity and even twenty minutes a day of movement, says Dr. Booth, is a decreased "prevalence of mortality and many chronic health conditions."

2. You Are Not a Number.

Scientists have created genome "scores" that try to determine the likelihood that someone will be an elite athlete. The scores combine various genetic traits that have been associated with endurance, speed, or strength. In theory, the higher your genomic score—the more of the genetic components your DNA carries—the more physically talented you should be. But in tests on actual Olympians, there was little correspondence between the genetic scores and real-world success. Some of the most successful athletes harbored few of the supposedly important genes.

3. Practice. Practice. Practice.

Multiple studies suggest that the key to success in many physical activities is training. Among runners, those with the highest training volume typically produce the lowest finishing times. Practice probably can outweigh a meager genetic endowment for athletes.

4. Don't Swab Your Child.

At the moment, several companies offer mail-order genetic testing that purports to be able to determine whether someone, usually a child, is destined for greatness in certain sports. The tests are crude in the extreme, usually scanning for only one gene, when hundreds and even thousands or tens of thousands of genes and portions of genes are involved in athletic success.

5. Do Steer Your Child Toward Fitness Early.

Really early. If a mother works out moderately during pregnancy, so, apparently, does her unborn child, which might affect physiology and gene expression throughout his or her life. Recent studies have shown that babies born to mothers who exercised had healthier hearts than other newborns. And in a touching experiment from Germany, when mothers-to-be were prompted to breathe fast and hard, as they would during exercise, their unborn children's hearts oscillated in response, synchronizing themselves, beat for beat, with their mothers'. Hormones could be the cause, the researchers noted. But they preferred to imagine a kind of music of the blood. The gasping breaths drove up the mothers' heartbeats, they wrote, until, inside the body, the sound grew loud, insistent, propulsive, and irresistible. The fetuses' hearts responded, the scientists hypothesized, settling into the same rhythm, with effects both physiological and poetic. A "pregnant mother's special awareness to the unborn child" may "be reflected by fetal-maternal interaction of cardiac activity," the German researchers concluded. As a mother runs and remakes her heart, the child she carries does the same.

10

Pushing Back the Finish Line

Not long ago, researchers affiliated with the Department of Twin Research and Genetic Epidemiology at King's College London discovered something surprising about some of the twins they studied. The pairs weren't aging at the same rate. Even some of the identical twins were aging differently.

Twins are interesting. I'm one, so I know. In most ways, we're like everyone else, but twins have that one other person who is particularly like them. Identical twins share their complete DNA. Fraternal twins share 50 percent of their DNA and, more important, a childhood and common early environment. This makes twins useful to science, and the UK Adult Twin Registry, the database that the King's College London researchers used, has been a gold mine of scientific information about lifestyles and health for decades.

But few of the other studies have produced results quite as striking as these.

For the experiment, the researchers contacted more than twenty-four thousand of the registered twins, with most of them fraternal and about three hundred identical. The scientists asked the twins about their lives, health, weight, what they did in their spare time, and how much they moved around during a typical week. Specifically, they had each twin fill out a detailed questionnaire about his or her activities and exercise regimen over the course of the previous year. The participants rated their activity level on a scale from 1 (slothful) to 4 (engaging in heavy activity). The twins also provided information about precisely how much time each week they spent on their various pursuits, whether sports, brisk walking, telly watching, or whatever, and how much time they had spent on those or similar activities when they were in their late teens or early twenties. At the time of the study, the twins ranged in age from eighteen to eighty-one, so for some, the "prior years" and current information were the same. For others, their most vigorous days were well in the past.

The researchers then drew blood from the twins and studied the state of their white blood cells. White blood cells serve many purposes in the body, primarily involving immunity. The body needs a large and steady supply of these cells, so they are constantly dividing, reproducing, and being reabsorbed into the moving tide of the blood. The health of a person's white blood cells can be a marker of general well-being.

In this case, after examining the cells microscopically, the researchers found that the most active pairs of twins had the youngest and most robust white blood cells. Their cells contained longer telomeres than those of less active twins.

Telomeres are the minuscule, protective caps on the ends of DNA strands, often compared to the tips of shoelaces and serving about the same purpose, to prevent fraying and tattering. Every time a cell divides, its DNA copies itself, but for reasons that aren't fully understood, the copying mechanism doesn't read all the way to the end of the DNA strand; instead, it slices off a tiny section, like cropping someone's scalp in a photo. This would be harmful, meaning genetic information was being lost, except that the snipping occurs in the telomere, which contains no genetic information.

The system works elegantly, until a telomere becomes too short, at which point DNA is compromised and the affected cell dies or enters permanent senescence, a kind of suspended animation. Most researchers accept telomere length as a reliable marker of cell age. The shorter a cell's telomeres, the more elderly and enfeebled the cell.

In the twin study, active people had the best-preserved telomeres. Those twins who exercised on average thirty minutes per day (which, in this study, represented "heavy activity") had telomeres as long and robust as sedentary people ten years younger. This difference remained even when researchers adjusted for body mass, gender, smoking, and other variables. Similarly, the twin pairs who'd been active in their twenties, no matter what their activity levels now, had longer telomeres than people who'd been sluggards all their lives.

Most interesting, though, were the twin pairs whose exercise routines had, as the scientists said, "diverged." One of the twins was active; the other was not. (There were few of these pairs, by the way, which tells you something about the tug of DNA and upbringing.) Presumably the twins had had almost identical telomere lengths at birth. But after years of either exercise or sloth, their cells had aged differently. The most active twin in almost every case had longer telomeres than his or her indolent sibling. Exercise had made these twins, the researchers said, "biologically younger."

The Myth of Birthdays

What is aging? That issue is surprisingly unsettled, in both science and society. Age, as most of us think of the thing, is not dependent solely on chronology. It's not just a matter of how many birthdays you've accumulated (or decided to skip; my twin sister and I, for instance, have reached an agreement whereby each of us acknowledges a new birthday only on alternate years, which has biological validity, because we did, after all, both agree). All of us know older people who glow with vitality and fill us with resentment, since they look better at fifty than we do at, um, thirty, and are probably having more sex, to boot.

But at the same time, age is not, as some people would claim, strictly a state of mind, either. It does demonstrably occur for most of us in the body and with some ferocity. The vast majority of people, as observation and science make clear, experience a rather precipitate physical decline at some point, usually beginning in middle age. We start to lose our muscle tone, waistline, height, energy, hair, motivation, sex drive, and car keys.

But what separates the relatively ageless from the rest of us? And can aging, as most of us experience it, be altered or slowed?

What, in other words, *is* aging, and how much do we influence it?

Those were among the questions that motivated McMaster pediatrics professor Dr. Mark Tarnopolsky, who undertook a series of recent experiments on whether aging is to some extent a lifestyle choice and if so, whether its effects can be ameliorated with a few (literal) steps.

To start, Dr. Tarnopolsky and his colleagues gathered a group of mice that had been bred to age at a hugely accelerated rate. Wild rodents as a rule have nasty, brutish, and short lives, but even by mouse standards, these mice had truncated life spans. They zipped through childhood and adolescence within weeks, enjoyed a few brief months of adulthood, and in most instances died of extreme physical decrepitude before their first birthdays.

Their accelerated aging was due to a genetic mutation that affected how well their bodies could repair malfunctioning mitochondria. Unlike other organelles within a cell, mitochondria have their own DNA, distinct from the cell's, and they can divide and multiply on their own. But in the process, they can accumulate small mutations, which normally are fixed by tiny, specialized repair systems within the cell. Over time, though, as we grow older, the number of mutations can outstrip the system's ability to make repairs, and mitochondria start malfunctioning and dying.

Some scientists believe that this process is the primary cause of aging. Not everyone agrees, though. Some researchers are convinced that bodily aging is caused by molecular attacks from free radicals (known as the oxidative stress theory of aging) or by simply eating too much (which causes a complicated cascade of other cellular problems). Those scientists who embrace the overeating theory of aging suggest, in turn, that dramatically cutting back on food will slow physical and mental decline. In other words, you starve yourself to longevity. This theory actually has worked in experiments with yeast cells and nematode worms, who live long, dull lives if you deprive them of most nutrients. But the same approach has proven to be wildly unpopular and inconclusive in studies involving higher organisms, such as rodents, dogs, and people.

On the other hand, there seems little doubt that mitochondrial health is somehow associated with the symptoms of aging, especially in mammals (if less so in nematodes). As mitochondria age, gather genetic faults, and falter, the cells they fuel wither or die. When that happens, muscles and sexual organs shrink, brain volume drops, hair falls out or turns gray, and soon enough we are, in appearance and beneath the surface, old.

This regrettable progression doesn't typically begin, though, until our mitochondria have exhausted their ability to repair themselves, usually sometime in our late forties or so. For many of Dr. Tarnopolsky's experimental mice, however, the problems began much earlier. They first developed malfunctioning mitochondria when they were as young as three months of age, or about age twenty in human terms. By the time they were eight months old, or about in their early sixties for us, the animals were extremely frail, with spindly muscles, shrunken brains, enlarged hearts, shriveled gonads, and patchy, graying fur, like rodent versions of Mr. Burns, if that isn't a redundancy. Listless, the animals barely shuffled around their cages, possibly muttering to themselves about how much nicer cages had been back when they were youngsters.

Every one of these mice died before reaching a year of age.

Except the mice that exercised.

Half of the genetically altered mice had been allowed to run on a wheel for forty-five minutes three times a week, beginning at three months. These rodent runners had been required to maintain a fairly brisk pace, Dr. Tarnopolsky says. "It was about like a person running a fifty- or fifty-five-minute ten-K," or covering 6.2 miles at about a nine-minute-mile pace. The mice continued this regimen for five months.

At the age of eight months, by which time their sedentary lab mates were bald, feeble, tottering, and dying, the running animals appeared vibrant and youthful, little furry Betty Whites. They had full pelts of dark fur—no salt-and-pepper shadings. They also had maintained almost all of their muscle mass and brain volume. Their gonads were the same size as when the mice had been young, as were their hearts. While the aged sedentary mice could barely stand without wobbling, the exercised mice balanced easily on narrow rods, the show-offs.

But perhaps most remarkable, although they still harbored the mutation that should have affected mitochondrial repair, they had more mitochondria overall and far fewer mutated mitochondria than the sedentary mice. At one year of age, none of the exercising mice had died of natural causes.

"We were surprised, to put it mildly, at how pervasive the effects were," Dr. Tarnopolsky told me. Exercise affected the aging process in every tissue and bodily system that the researchers studied. Even the younger researchers were impressed. While Dr. Tarnopolsky, an athlete for most of his life, was relieved to see that the active aged mice had kept their hair, his graduate students were more concerned with the animals' robust gonads. Their testicles and ovaries hadn't wrinkled or shrunk, unlike those in the inactive elderly mice.

"I think all of my researchers exercise now," he said.

The Great Shrinking

It's important to remember that not long ago most people, including scientists, were convinced that physical aging was a slow but inevitable march to frailty, and it was best if we took to our rockers sooner rather than later to protect our fragile bones and egos. This attitude was understandable. In the 1970s, a number of major studies had indicated that beginning as early as people's late thirties, they start to lose muscle mass, a process known as sarcopenia. Since the studies found signs of sarcopenia in all of the subjects, the researchers concluded that an accelerating loss was inevitable. If you lived to middle age, you'd have less muscle year after year, until your total muscle mass was declining at least 1 percent per year, every year. Strength plummeted in lockstep.

Sarcopenia is widely recognized as one of the primary causes of frailty in older people and the accompanying loss of independence. Someone who is puny and whose muscles have shriveled can't rise from a chair or heft grocery bags. He or she can't live alone.

Disquieting follow-up studies found that in many elderly people, even the muscle tissue that remained was ailing at a molecular level. It contained fewer satellite cells than in muscle tissue from young people. Satellite cells are an essential, specialized type of stem cell that regenerates muscle tissue. Without sufficient satellite cells, muscle can't rebuild and strengthen itself. It becomes prone to tears and strains.

Still other studies found that the muscles of people past age forty also showed signs of reduced mitochondrial activity. Mito-

chondria, as Dr. Tarnopolsky would remind us, are the power en-
gines of cells; they convert food fuel into forms that can be used by
the muscles. Without enough healthy mitochondria, muscles become weak and easily exhausted.

Meanwhile, other body parts and systems also sagged, thinned,
or failed under the onslaught of years, as science helpfully pointed
out for us. Multiple large epidemiological studies indicated that
bones, in particular, lost thickness and grew brittle in both men and
women after middle age. As with muscles, the quality of a person's
bones determines, to a very large degree, the quality of life. Just try
standing upright without a functioning skeleton. Go on, try.

Scientists found, too, that people's general physical fitness, as
determined by the ability of their respiratory and circulatory systems
to deliver oxygen to laboring muscles, fell by about 10 percent per
decade after age forty. That loss represents a considerable dip in lung
power and is a primary reason, many researchers think, why many
middle-aged people reduce exercise or quit altogether. The effort
feels increasingly hard.

In essence, science concluded, older people had lousier-quality
muscles, bones, and physical ability than the young (who, of course,
don't appreciate what they have; they never do).

But there was a signal failing in all of these studies: They relied
almost exclusively on volunteers who were inactive. (Most Americans are; it's easy to find and study those folks.)

Everything changed when scientists began looking at volunteers who willingly and frequently moved.

Grandpa Will Pass You Now

A recent study of participation in the New York City Marathon, the
nation's largest, found that throughout the early 1980s, the number
of registered runners in the fifty-plus category remained negligible,

even as participation soared among younger runners. At the time, the authors of the study wrote, "few 60-year-old men, much less women—or their doctors—would have considered it possible for someone of that age to run 26-plus miles."

My, how times change. While the number of male racers in their 20s at the New York City marathon grew by about 25 percent from the 1980s to the 2000s, the number in their 50s jumped by a remarkable 78 percent, and participation by female masters marathoners swelled by more than 30 percent. And these well-seasoned runners were fast. During those years, the average finishing time for younger runners increased by almost 30 minutes—meaning the kids were getting slower. But the typical finishing time for older runners fell. The 70-plus men shaved nearly two minutes from their average time, and the 60-plus women dropped their average finishing time by twice that.

Even more remarkably, during those same years, masters participation in the grueling Hawaii Ironman began to climb. The Ironman race, as you may know, consists of a 2.4-mile ocean swim, 112 miles of cycling, *and* a full marathon. Far more physically taxing than a mere marathon, the race had, at its inception three decades ago, no category for racers past sixty. Now that age group and beyond are well represented, especially among men, according to an analysis of the event published recently in *Medicine & Science in Sports & Exercise*. A few ninety-year-olds have covered the distance. These older triathletes, the authors of the Ironman study wrote, "represent a fascinating model of exceptionally successful aging."

They represent even more a repudiation of much that past science and convention have told us about aging.

Or as Dr. Hirofumi Tanaka, a professor of physiology at the University of Texas at Austin and an expert on aging athletes, says, "A great deal of the physical effects that we once thought were caused by aging are actually the result of inactivity."

And that, ladies and gentlemen, we can change.

The impacts can be remarkable. Let's look, for instance, at your satellite cells, those specialized cells that help to repair and regenerate muscle tissue. A recent experiment with aged, inactive rats showed that, as expected, the old animals' leg muscles contained far fewer restorative satellite cells than those of young rats. But when the elderly animals were given access to running wheels, their leg muscles began sprouting new, hardy populations of satellite cells, suggesting that their muscles would now be able to build and repair themselves effectively.

Dr. Tarnopolsky and some of his colleagues discovered similar changes in the leg muscles of active older people. They contained far more satellite cells than the muscles of inactive people of the same age.

Other aspects of mature muscles also are affected by movement. Recently, researchers from the Canadian Centre for Activity and Aging reported that when they microscopically examined the leg muscles of older human runners, most of them past age sixty-five, they found that the aging runners' muscles were densely packed with motor units. A motor unit is, essentially, the control mechanism of a functioning muscle, composed of a neuron and the particular muscle fibers that that neuron activates. The more motor units in a muscle, the faster and more fully that muscle can contract; the healthier and stronger it is. Too few motor units indicate poor muscle tissue health and possibly the beginnings of sarcopenia.

No such problems plagued the sexagenarian runners. Their leg muscles teemed with almost as many motor units as in active twenty-five-year-olds. Running, the scientists wrote, seemed able to "potentially mitigate the loss of motor units with aging well into the seventh decade of life."

It seems to have the same effect on bones. Several studies have found that active older men and especially active older women have healthier, thicker bones than their compatriots who sweat less.

And, of course, there is the impact of exercise on telomeres, which is arguably the most pervasive antiaging effect. In proof of that, scientists in Germany rounded up men and women, some young, some middle-aged, all sedentary, and then went after the active. They recruited a set of professional runners in their twenties, most of them on the national track-and-field team, and a separate group of serious middle-aged longtime runners, with an average age of fifty-one and an impressive training regimen of about fifty miles per week.

From the outset, the scientists were struck by one particular aspect of their older runners. "They looked much younger than sedentary control subjects of the same age," says Dr. Christian Werner, an internal medicine specialist and the study's lead author.

Even more striking was what was going on beneath the deceptively youthful surfaces. When the scientists examined white blood cells from the young adults, they found that all of them harbored similar-size telomeres, whether they were runners or sedentary, a finding the scientists had anticipated. None of the young people had been on earth long enough, after all, for multiple cell divisions to have snipped away at their telomeres.

But in the middle-aged the situation was quite different. The sedentary midlife subjects had telomeres that were about 40 percent shorter than in the sedentary young people, suggesting that the older subjects' cells were, like they were, aging. The runners, on the other hand, had remarkably youthful telomeres, a bit shorter than those in the twentysomething runners, but only by about 10 percent. In general, telomere loss was reduced by approximately 75 percent in the aging runners.

Exercise, Dr. Werner says, "at the molecular level has a strong antiaging effect."

"These Experiments Changed How I Work Out"

Just how exercise exerts its youthful-izing effects remains largely unknown and is probably, Dr. Tarnopolsky says, "complicated, multisystemic, and multifactorial." In his mouse experiments, for instance, running resulted in an upsurge in the rodents' production of a protein known as PGC-1alpha, which regulates genes involved in metabolism and energy creation, including mitochondrial function. Exercise also sparked the repair of malfunctioning mitochondria through a mechanism outside the known repair pathway; in these mutant mice, that pathway didn't exist, but their mitochondria were nonetheless being repaired.

Other mechanisms seem to be at work in the influence of exercise on aging bones, and they extend deep below the surface. In seminal, recent experiments at the University of North Carolina, researchers removed stem cells from the bone marrow of animals and cultured them. The cells in question, mesenchymal stem cells, are fairly specialized. They typically transform into either bone or fat cells (and far less commonly, other tissues).

After any stem cell differentiates, of course, it can't be anything else: Once a fat cell, always a fat cell; once a bone cell, etc. Any stem cell that becomes fat, though, is a stem cell that did not become bone, and as a result, that portion of your skeleton to which a bone cell might have migrated receives no new cell. It will be that littlest bit punier.

Unfortunately, stem cells like to become fat. In most of the North Carolina scientists' experiments, that's what the stem cells did. "They love to become fat cells," says Dr. Janet Rubin, a professor of medicine who led the studies. "It's discouragingly easy to nudge them in that direction."

But that changed when the cells were, in effect, exercised. When the petri dishes were stimulated with high-magnitude mechanical signals, similar to the force that moves through your leg bones when you go for a jog, most of the stem cells did not become fat cells. "There was a really striking difference in outcomes," Dr. Rubin says. When the cells weren't allowed indolently to loll about in their petri dishes but were instead put through a cellular workout, they did not become fat.

The lesson, she told me, is simple. Movement is necessary to stimulate the biochemical signals that direct stem cells to become bone.

"This is the first time in my career that something I've done in the lab has changed how I work out," she added. "I'm in my fifties. I wish to have healthy bones for the next five decades." So you'll find her on the treadmill almost every day and twice a day on weekends.

What You Can't Outrun

There are limits, of course, to what exercise can accomplish against the onslaught of the years. Only in fantasy and among very hungry nematodes is extreme life extension possible. "Aging is tricky," says the Mayo Clinic's Dr. Michael Joyner, an accomplished triathlete in his 60s, as well as an expert on the physiology of older masters athletes. "Most of us can do a lot more than we think we can. But we also have to accept that some things can't be changed."

Flexibility, for example, is reduced in virtually everyone who's reached the age of fifty, compared with their younger selves, most studies show. People who once could wrap their feet behind their ears, to the delight of their significant others, find that they now have trouble touching their toes, and those who never could touch their toes bend these days with the grace and ease of an ironing board.

Even being active won't slow this loss of limberness, most science suggests. A lamentable recent study of older men and women found that flexibility in the hips and shoulders was about 6 percent lower with every additional decade after fifty, even if people regularly exercised.

Why joints become less flexible with age is controversial, but many researchers believe that the collagen in aging connective tissue degrades. This process probably contributes to a fairly high risk of injury in middle-aged competitive athletes. "People tell me all the time, 'I never got hurt when I was young, and now things ache all of the time,'" Dr. Joyner says. In epidemiological studies of competitive runners, the injury rate, already high, rises after age fifty, even among lifelong athletes who rarely had experienced injuries before.

Interestingly, though, the incidence of knee arthritis is not necessarily high. As I mentioned earlier, a famous Stanford University study followed middle-aged longtime runners for two decades, beginning when most were in their fifties or sixties. After twenty years, the runners' knees were noticeably healthier than non-runners'.

Still, the lesson of healthy aging is that for those of us who plan to be active into our second century, a certain amount of care is required. "There is very little evidence that stretching, by itself," and especially just before exercise, improves flexibility or reduces injuries in middle-aged and older athletes, Dr. Joyner says. But more active forms of stretching, such as moving your shoulders through an entire service stroke multiple times before starting a tennis match, are "probably advisable," he says.

It's also conceivable that nature intends for those of us who have been physically hyperactive for most of our lives to slow down at least somewhat as we age. Provocative studies with mice have shown that in strains bred to love running, animals will spontaneously begin to reduce their daily mileage when they reach rodent middle age. As youngsters, they might have skittered on their wheels for six hours straight. Now they'll jog sedately for two hours or so, then quietly contemplate the far side of their cage for a while.

The same process may occur, to some degree, in people. "We

have yet to see anyone in our studies who maintained the same training volume and quality" at the age of 60, 70, or 80 as at the age of 30, Dr. Tanaka says. Other researchers have found that most 50-plus athletes, even the most committed, train about half as much as 20- and 30-year-old athletes. "It may be that we're all really busy at that age," Dr. Tanaka says. "But it may also be that we're programmed to do a little bit less as we get older. The key point is to keep doing something."

And for those who require positive reinforcement, the science unequivocally shows that you can return to activity, with considerable benefits, even if you've been inactive for years. An interesting recent study of how unexercised bodies, young and old, respond to rigorous endurance training found that young people (average age twenty-four) became much more physically fit after two months of intense interval sessions on stationary bicycles. Their average VO2 max rose by about 13 percent.

But the older volunteers (average age sixty) also improved their endurance significantly. Although their average VO2 max rose by only about 8 percent, it did rise. This is important not just for bragging rights. Aerobic fitness has other measurable and ineffable effects as we age. Numerous studies of mortality and health have found that for every percentage rise in someone's aerobic fitness, their risk of premature death falls. In one study of almost fifteen thousand European men, those with the highest VO2 max had a 50 percent lower risk of premature death than the men with the poorest VO2 max numbers. In another, American study, involving more than twenty-six hundred men and women, aerobic fitness was a better predictor of longevity than any other health measurement the researchers looked at, including waist circumference, smoking history, and body mass index. Even obese participants and smokers lived longer if they were aerobically fit. So those newly fit sexagenarians, with their 8 percent VO2 max increase, conceivably bought themselves another decade of life.

"Mysteries still abound," Dr. Tarnopolsky says, about why cells age and what impacts activity can have on the process. But one message is unambiguous. "Exercise alters the course of aging."

Dr. Joyner agrees. The really great news in the latest science about exercise and aging, he says, "is that it shows that aging is, to some degree, adjustable. You can make lifestyle choices that directly affect how well you age." Walk, run, or triathlon (if that is a verb). Jiggle your stem cells. Stretch your telomeres. Boost your satellite cell count. "If you're active," Dr. Joyner concludes, "aging will not be a slow march to frailty." It will be a conga dance to a far-off finish line.

A Few Simple Steps Can Change How You Age

1. Even a Little Exercise Is Better Than None.

A 2008 international study of exercise and cellular aging concluded that "moderate physical activity levels may provide a protective effect." You don't have to exhaust yourself, in other words, to potentially protect your cells.

"There is probably a threshold amount of exercise" that is necessary to affect physiological aging, says Dr. Tarnopolsky, who's extensively studied aging and exercise, "but anything is better than nothing." If you haven't been active in the past, he continues, "start working out for five minutes a day, then increase your activity level five minutes at a time."

2. Take Actual Steps.

Walking is a wonderful exercise, "especially if your main goal is good health and to lessen the effects of aging," says Dr. Joyner. As proof, he points to the work of Hiroshi Nose, M.D., Ph.D., a professor of sports medical sciences at Shinshu University Graduate School of Medicine, in Japan, who has enrolled thousands of older Japanese citizens in an innovative, five-month-long program of brisk, interval-style walking (three minutes of fast walking followed by three minutes of slower walking, repeated ten times). The results have been notable, particularly in terms of aging. "Physical fitness (maximal aerobic power and thigh muscle strength) increased by about 20 percent," Dr. Nose wrote to me in an e-mail, "which is sure to make you feel about 10 years younger than before training."

3. For Competitive Older Athletes, Two Words: Intervals. Sorry.

"The main thing that we see, from a scientific standpoint, is that if you want to maintain your fitness and performance levels as you age, you have to practice intensively," says Dr. Tanaka, who has extensively studied masters athletes. Intensive exercise affects the heart and lungs in ways that keep your VO2 max as high as possible. In practice, this means intervals, or short, hard, repeated bursts of exercise, such as sprinting for a quarter of a mile, resting for a few minutes, then sprinting again. "I don't know many athletes of any age who love intervals," says Dr. Tanaka. "But they are important if you want to compete as you get older."

4. Resist.

Grip strength is another of the most reliable predictors of quality of life as we get older. How well can you hold things and crush someone's hand in yours to establish dominance? But strength of all kinds does tend to diminish, especially if you're inactive. When Canadian scientists studied the leg and biceps muscles of longtime middle-aged runners, they found that their leg muscles teemed with healthy motor units, which allow the muscles to contract quickly and with force. But their arm muscles were puny, having far fewer motor units than those of younger people. They weren't using their arms much to run.

Other research has found that resistance training in people over the age of sixty increases joint range of motion and reduces injuries. Interestingly, these benefits occur even though most older athletes, especially men, don't add much bulk as a result of weight training. A review of dozens of studies of weight training in older people found that, in general, any amount and type of weight training improved older people's body compositions; they had less fat and more lean tissue afterward, gaining typically about 1 kilogram or 2.2 pounds of muscle, which may "seem modest compared to the expected adaptation with healthy young people," the authors admitted, but it's a much better outcome than "the 0.18 kilogram" or .4 pounds "annual decline" in muscle that occurs "with sedentary lifestyles beyond 50 years of age." You can find strength training advice and routines at the end of chapter 6. There are no indications that those of us who are no longer in the first blush of youth (and thank goodness; remember how embarrassing those first blushes were?) need to back off significantly from weight training regimens aimed at younger people. Progression is key, and keeping it is, too, the review concluded. Another study found that older men tended to lose their strength gains faster than thirtysomething athletes when both groups quit weight training for several weeks. But after returning to the gym, they soon regained the lost strength. Jack LaLanne, after all, never complained that he could no longer tow a caboose or complete 100 one-arm push-ups merely because he'd turned 70. Or 80. Or, in fact, 90. A model worth emulating, apart, please, from the skintight orange bodysuit.

Conclusion

Use It or Lose It

Each organic being is striving ... and the vigorous, the healthy, and the happy survive and multiply.

—Charles Darwin, *On the Origin of Species*

As far back as the early 1700s, the Italian physician Bernardino Ramazzini noticed that "chair workers," men such as cobblers and tailors, whose professions constrained them to sit for long periods of time, were less healthy than the low-paid dogsbodies who fetched fabric and ran errands for them.

This observation, though provocative, was ignored for centuries, until the British physician and epidemiologist Jeremy Morris began studying bus conductors. Starting in 1949, Dr. Morris and his colleagues gathered health and occupational data about conductors and drivers for London's iconic double-decker buses. The researchers quickly determined that the drivers sat for about 90 percent of each workday, while the conductors walked and climbed stairs constantly, ascending and descending as many as six hundred steps during each shift. Using medical records supplied by London's transportation agency, Dr. Morris cross-correlated the figures and found that the sedentary drivers as a group were more than twice as likely to suffer a heart attack as the more active conductors. This risk held steady even when the researchers considered an unusual, additional data set from the transportation agency detailing the size of the trousers that the bureau supplied to its workers. In general, the conductors maintained narrower waists than the drivers. But even among conductors with a Falstaffian physique, heart attacks were much less common than among drivers, thin or not.

Dr. Morris's work, famous among physiologists, provided some of the first hard data showing that movement is healthy, particularly for the heart, and that sitting is not. To validate the finding, Dr. Morris conducted a follow-up examination of British postal workers, and, as with the bus employees, found that mail carriers who walked or bicycled along their delivery routes were significantly less likely to develop or die from heart disease than postal clerks and telephone operators, who sat quietly back at the post office all day.

With this persuasive evidence of the robust health benefits of activity to guide us, Americans and the British became, in the half-century or so after these studies were published, the most sedentary group of humans ever to exist, a category that includes many of us who exercise regularly.

Active Couch Potatoes

In 1982, researchers affiliated with the Cooper Institute surveyed a large group of well-educated, affluent men. The researchers were interested in the men's exercise habits, but they also asked, almost incidentally, about their indolence. Specifically, they inquired about how many hours each day the men spent watching television or sitting in a car. (This was before you could do both at once.) Over the years, the survey's main results were used to bolster the idea, already emphasized by Dr. Morris, that exercise is healthy.

But no one had really looked at the residual information about how the men spent their time when they weren't exercising. Then, a few years ago, scientists from the University of South Carolina and the Pennington Biomedical Research Center parsed the full data.

To no one's surprise, they found that, as Dr. Morris had noticed with bus drivers, the people who sat the most had a heightened risk of heart problems. In this case, the men who spent more than twenty-three hours a week watching TV and sitting in their cars (as passengers or as drivers) had a 64 percent greater chance of dying from heart disease than those who sat for eleven hours a week or less. What was unexpected was that many of the men who sat long hours and developed heart problems also exercised. Quite a few of them said they did so regularly and led active lifestyles. The men worked out, and then sat in cars and in front of televisions for hours, and their risk of heart disease soared, despite the exercise. Their workouts did not counteract the ill effects of sitting.

Most of us know that being sedentary is unhealthy. But many of us simultaneously think that being sedentary is something that happens to other people. We spend our lunch hours conscientiously jogging or power walking or visiting the gym. But then we drive back to the office, settle at our desks, and sit for the rest of the day. We are, in the parlance of concerned physiologists, "active couch potatoes."

"This is a very new occurrence," says Dr. Booth, an expert on inactivity, of this pattern of modern life in which someone dutifully exercises for a certain period of time but is physically inert for much of the rest of the day. The pattern is widespread, though. The amount of time that most Americans, even those who exercise, spend being otherwise inactive has risen steadily in recent decades. According to new estimates, most of us spend more than nine hours a day in oxymoronic "sedentary activities."

Pervasive changes in how we work have contributed to our growing sluggishness. Until the 1960s or so, a wide-ranging recent survey of workplace environments concluded, a majority of occupations in the United States involved moderate physical activity. Now, 80 percent of jobs in this country are almost completely sedentary, the survey found. We sit at screens. We eat at our desks. We don't even walk down the hall to gossip; we text our coworkers.

Commutes and cable have also done much to reduce our movement. In the Cooper Institute study from the 1980s, many of the men spent three hours or more each day in the car or in front of the television, numbers that now seem quaintly low. In more recent studies of daily behavior, researchers have found that Americans regularly slump for five hours or more daily in front of the television, and the screen's pull begins early: Toddlers often watch for four to six hours a day. Television viewing serves as a handy simulacrum of general inactivity, since it is easily measurable and reliably slothful. We don't walk in place while watching *The Voice*. We sit.

At the same time, we don't work around the house as much anymore. Decades ago, before the advent of computers, Chinese takeout, Roombas, and, of course, fifty-four-inch plasma TVs, people occupied themselves for hours (not always enthusiastically) with "light-intensity activities," as physiologists define them, meaning actions that require you to move around the room but not break a sweat. Mopping, gardening, vacuuming, cooking, hedge pruning, chicken slaughtering and plucking, and changing lightbulbs all qualify. Few of us accumulate much "light-intensity activity" nowadays. We've replaced those hours with sitting.

And the physiological consequences are considerable. In the largest study to date of current activity patterns, scientists with the National Cancer Institute spent eight years following almost 250,000 American adults aged fifty to seventy-one. When the study began, the participants answered a series of detailed questions about how much time they spent commuting, watching TV, sitting before a computer, and exercising, as well as about general health information. At the time, none of the volunteers suffered from heart disease, cancer, or diabetes. After eight years, though, many were ill and quite a few had died. These unhealthy (or worse) respondents tended to be the most sedentary. Those who watched TV for seven or more hours a day were at greater risk for all-cause, cardiovascular, and cancer mortality, the authors wrote. And isolated bouts of exercise did not do much to ameliorate the risk. People who exercised for seven hours or more a week but spent at least five hours a day in front of the television were more likely to die prematurely than the small group who worked out seven hours a week and watched less than an hour of TV a day.

If those numbers seem abstract, a new Australian study provides some essential and unsettling grounding. In it, researchers determined that for any adult, including us, inactivity reduces life span. Every hour of television that a person watches after the age of twenty-five, the researchers concluded, potentially snips twenty-two minutes off of the viewer's life span. If an average man watched no TV in his adult life, his life span might be 1.8 years longer, the scientists wrote, and a TV-less woman might live for a year and a half longer than otherwise.

Our Cells Don't Like Our Lifestyle

Just how sitting adversely affects the health of even the well-exercised is an issue only slowly now being untangled by scientists. But it's clear that the answer involves, to some extent, muscle contractions. When you stand, even if you don't walk around, the large muscles in your back, buttocks, and legs contract to keep you upright and stable, to keep you from wobbling. When you sit, those muscles aren't needed.

In animal studies, when rats or mice were unable to put weight on their legs, which were either lifted off the ground with a miniature traction device or set in plaster casts, the animals rapidly developed unhealthy cellular changes throughout their bodies. In particular, they produced substantially less lipoprotein lipase, an enzyme that is known to aid in the breakdown of fat in the bloodstream. Without sufficient levels of the enzyme, fat accumulates in the blood, migrates to the heart, and can jump-start cardiovascular disease.

Tests in humans have found even more pervasive impacts. A review of studies in which people volunteered to be inactive found that "13 different cardiometabolic parameters" deteriorated in the volunteers within four months. Insulin sensitivity is especially easy to disrupt. When a group of healthy young men were confined to their beds in a particular study and not allowed to place weight on their legs for any reason (they were wheeled to the bathroom as needed), they developed symptoms of insulin resistance within twenty-four hours; their slumberous muscles were 40 percent less capable of slurping glucose or sugar out of the bloodstream.

Even less drastic reductions in movement produce sizable and rapid consequences, other studies have shown. In one, twelve bloomingly healthy young adults who had been taking, on average, ten thousand or more steps a day based on pedometer readings were asked to cut back to less than forty-five hundred steps a day. They spent that spare time sitting. After three days, their levels of blood sugar after meals had risen dramatically, by as much as 90 percent, and their insulin response had slowed, early symptoms of insulin problems and eventual diabetes.

Perhaps most disturbing of all, scientists have found that occasional bouts of endurance exercise do not fully return all of these disrupted bodily systems to normal. Lipoprotein lipase levels, for instance, seem to be regulated by how much you sit, not by whether you sweat. "There seem to be different pathways" involved in the beneficial physiological effects of exercising and the damaging impacts of sitting, says Tatiana Warren, Ph.D., the lead author of the study of men who sat too much. "One does not undo the other," she says.

Or, as Dr. Booth has written, "Inactivity is not just not getting exercise. It has its own physiology."

And the impacts run deep. A wrenching new study of five- and six-year-old children who were inactive and overweight found signs of incipient "injury" to certain areas of their DNA, according to the researchers. Active children's DNA showed no similar damage.

"Human cells are maladapted to an inactive lifestyle," Dr. Booth says. In other words, we are not designed to be still.

Move It!

So this is a call to arms—and legs, muscles, and lungs—as well as naked self-interest. Each of us needs, almost certainly, to move more. Set aside this book and stand up. Your physiology will be different. Your future, if only by the slightest degree, will be changed.

Done? Back? Good. Keep bobbing up periodically. An inspiring study presented at a recent American College of Sports Medicine annual meeting found that this simple expedient—of standing up and moving about from time to time—improved people's health significantly. The scientists had a large group of adults either sit completely still for seven hours or, on a separate day, rise and walk leisurely in place for two minutes at twenty-minute intervals. On another day, they had the volunteers jog in place during their breaks. When the volunteers remained stationary for the full seven hours, their blood sugar spiked and insulin levels were out of whack. But when they broke up the hours with movement, even just a short two-minute stroll around the room, their blood sugar levels remained stable. Interestingly, jogging didn't improve blood sugar regulation any more than standing and walking did. What was important, the scientists concluded, was simply breaking up the long, interminable hours of sitting.

Still need convincing? A separate recent study showed that motion, even in diminutive, largely unintended increments, aids in weight control. When researchers compared a group of lean and obese women, none of whom formally exercised, they found that the leaner women tended to fidget. They wiggled and bounced in their chairs, and in general had difficulty sitting still. This fidgeting added up, in terms of energy use. "If the obese women adopted the activity patterns of the lean women," the authors wrote, they would burn an additional three hundred calories every day.

So move. Formally exercise, of course—there are benefits available from endurance and weight training that standing or fidgeting, on their own, cannot match. But stand up, too. I realize that if you have read this far, I am probably preaching to the choir. But reminders are never amiss. I need them. Few things, I think, impinge as much on an active life as writing about exercise—all that time before a computer or reading medical journals; those countless hours of feckless, seated procrastination. My lipoprotein lipase evaporated. Fat seeped insidiously into my blood, muscles, and ventricles; stupor infiltrated my brain.

I've taught myself, in defense, to stand up and pace when I'm on the phone, and I prop papers on a music stand, so that I can be upright while I read. (I stand on one foot when I brush my teeth at night, too, which has little to do with activity but may be one of the more transformative actions I've picked up from researching this book. My balance and physical confidence have improved noticeably, and my husband remains, even now, amused, which is not a bad foundation for marital health.)

We as humans are made to move. All of the available science and our own common sense make clear that "humans have an intrinsic biological requirement for a certain threshold of physical activity," as a recent editorial in the *Journal of Applied Physiology* pointed out. We must move our bodies or we begin to lose them, our internal systems eroding and our physical potential and capabilities slipping away. If we're sluggish when young, we lose years from our life span as we age. Even before that, we sacrifice full, independent, satisfying lives. We grow frail when we might have been hearty, and sick when we might have been well. "Inactivity is the greatest public health threat of this century," Dr. Booth says. "And it is almost completely preventable."

So, for the last time, don't take these tidings lying down. Get up, and pull your family and friends along with you. Go for a hike, bike ride, or run together. Pass this book along to your mother, her mother, your children, your general practitioner, and your children's soccer coach. Pass it back and forth between yourself and a friend twenty-five times or so and you will in fact have completed a light but adequate core- and balance-training session, and my work will, I feel, be done, at which point I will leave you to get on with your workout and head out for my own easy, life-altering run.

Index

nutrition *(cont.)*
  and dietary fads, 69
  of early humans, 210
  and exercise, 88–91
  fluids, 58–64, 75
  fructose, 56–58
  low-carb/high-fat diets, 88–91
  post-workout, 66–68, 76
  protein, 67–68, 76
  sports nutrition, 68–69
  supplements, 70–74
  trends in, 51, 69
  See also carbohydrates; weight and
    weight loss

one-repetition maximum, 142
overload principle, 9–10, 20–21
overtraining, 118–20
overuse injuries, 16, 23, 27–28, 156–59,
  169–72

pedometers, 97
perspiration, 151–53
Phillips, Shane, 89
Phillips, Stuart, 127, 130, 135, 136,
  137, 138, 142, 144–45
pickle juice, 154–56
Pilates, 143–44
plyometrics, 132, 137, 147
polyphenols, 73–74
power, 130–32
power athletes and genetic advantages,
  221
predisposition for exercise, 213–16
prenatal environments, 217–19, 229–30
protein, 67–68, 76
psychosomaticism, 44
push-ups, 143, 145

Ramazzini, Bernardino, 249
Rankinen, Tuomo, 214, 227–28
rating of perceived exertion (RPE), 112
Ravussin, Eric, 80, 82, 96
Reiff, Ralph, 118–19

resistance training, 128. *See also*
  strength training
resting, 44–46, 117–20
resting heart rate, 103, 106–7
Robling, Alexander G., 137
Roth, Stephen, 213
Rubin, Janet, 242–43
running
  and aging, 238–39, 240–41
  barefoot running, 160–62, 164
  beginning runners, 106
  and bone health, 137
  and core strength, 140
  efficiency of, xiii–xiv, 209–10
  elite runners, 105
  and fartlek, 108–9, 112
  as fitness indicator, 102–3, 121
  and fluids, 58–59, 60, 61–62, 63–64
  form in, 162–63, 164–65, 176–78
  genetic advantages in, 216
  and heart health, 102–3
  and ibuprofen, 38
  injuries from, 157–59, 160–62
  Kenyan runners, 58–59, 63, 217–18,
    219
  and knee health, 148–49, 173–74,
    244
  and nutrition, 73
  and overtraining, 118–19
  and power training, 130–31, 132
  and resting, 45, 118
  shoes for, 157–59, 177–78
  and stretching, 27–28, 43–44
  on a treadmill, 122
  volume of training in, 109–10, 222
  warming up for, 35, 47
Ryan, Michael, 159

Salazar, Alberto, 120
Scherr, Johannes, 73
Schiller, Friedrich, xiii
Schoene, Robert, 119–20
sedentary lifestyles
  and aging, 239